# Essentials of Oral Biology

CHURCHILL LIVINGSTONE DENTAL BOOKS

# Essentials of
# Oral Biology

## David Adams
B.Sc., M.D.S., Ph.D.,
Reader in Oral Biology,
Honorary Consultant Dental Surgeon,
Department of Oral Biology,
Welsh National School of Medicine,
Dental School,
Heath, Cardiff

CHURCHILL LIVINGSTONE
EDINBURGH LONDON MELBOURNE AND NEW YORK 1981

CHURCHILL LIVINGSTONE
Medical Division of Longman Group Limited

Distributed in the United States of America by
Churchill Livingstone Inc., 19 West 44th Street, New
York, N.Y. 10036, and by associated companies,
branches and representatives throughout the world.

First published 1981

ISBN 0 443 02095 7

**British Library Cataloguing in Publication Data**

Adams, David
  Essentials of oral biology.
  1.   Mouth
  2.   Teeth
  I.   Title
  612       RK280       80-41014

Printed in Singapore by Singapore Offset Printing Pte Ltd

# Preface

The aim of this little book is to guide the dental undergraduate
student to those aspects of the basic sciences which are of direct
relevance to his clinical work. It is assumed that he will already have
had instruction in the general aspects of these subjects. The book is,
therefore, unequivocally vocational in its treatment of the subject. It
is not meant to replace the more comprehensive texts, to which the
student will need to refer for details of some aspects that are only dealt
with in outline here. I have not attempted to include the very latest
information in dental research, as the book is not intended for the
postgraduate research worker. However, the graduate studying for
the Primary Fellowship examination of the Royal Colleges may find it
useful for revision.

There is no differentiation between anatomy, physiology and
biochemistry because I believe that these subjects should be
integrated at this stage of the course. In a sense this book is meant to
be the lowest common denominator of the various patterns of the
teaching, in dental schools, of the biology of the mouth.

Cardiff, 1980                                        D. A.

# Acknowledgements

Fortunately in writing a textbook there are many people who are willing to give assistance. I am grateful to Dr D. K. Whittaker and Professor D. Picton for helpful advice on reading parts of the manuscript, to the staff of the Audiovisual Aids Department at the Dental School in Cardiff for much of the illustrative material. Figures 4.4, 8.1, 8.2, 8.3, 8.6, 8.8, 8.10, 9.3, 9.5 are taken from *Introduction to Dental Anatomy* by Scott and Symons, and I thank Professor N. B. B. Symons for permission to reproduce them here. I also acknowledge the permission of Professor N. Jenkins and Blackwell Scientific Publications Limited to reproduce Figure 12.2 from the *Physiology and Biochemistry of the Mouth*. Figure 4.10 first appeared in *Applied Physiology of the Mouth*, edited by C. L. B. Lavelle and published by John Wright & Sons of Bristol. Thanks are especially due to Miss C. A. Edwards who typed the manuscript.

Finally my thanks are due to the publishers, Churchill Livingstone, who gave me invaluable advice and encouragement.

1981                                                     D. A.

# Contents

# 1

# Terminology

When embarking on a new subject terminology is one of the first difficulties a student encounters. The working language in the dental clinic will be foreign to the new student until he has acquired the basic vocabulary. To add to the confusion there is a tendency nowadays to communicate in a 'shorthand' style. Abbreviations are rife and it is not uncommon to hear and see sentences which are almost totally in initial letters. For example: 'RCT without ABC for a patient with CV disease may result in SBE' when translated reads, 'Root canal therapy without antibiotic cover for a patient with cardio-vascular disease may result in sub-acute bacterial endocarditis'.

Glossaries are usually placed at the end of textbooks but as I believe that the sooner the vocabulary is learnt the better the understanding the student will have of the subject, I have placed it at the beginning. Some of the terms presented here will be defined again later and in a more specific manner, but my aim is to provide a working vocabulary that will be enlarged as time goes on and more knowledge is gained. Thus, the terms selected for presentation are those which are in common usage in the field of oral biology.

## The mouth and teeth

*Oral* is an adjective which is almost synonymous with the mouth. It should not be confused with aural which relates to the ear. Oral Biology means the study of the structure and functions of the tissues within and around the oral cavity. Another word denoting the mouth is *stoma*. It is the stem from which the word stomatology comes. Stomatitis is also from this stem and means inflammation of the mucosa of the mouth.

*Buccal* refers to the cheeks and this word like oral is an adjective. Buccinator muscle or cheek muscle obviously has the same source. The adjective is used to describe position, i.e., the buccal aspect or buccal side of a tooth is the side facing the cheeks. The buccal sulcus is the trough between the teeth and the cheeks.

*Labial* refers to the lips in the same way as buccal refers to the cheeks. We talk of the labial side of the anterior teeth as that side which is towards the lips.

*Lingual* means pertaining to the tongue and again the word is often used to describe position, e.g., the lingual side of the tooth faces towards the tongue as opposed to the buccal side. It is used when referring to structures of the lower jaw since when we refer to the structures of the upper jaw the word *palatal* denotes the inner surface. The other term related to the tongue *'glosso'* is a Greek word and appears as a prefix to some words, e.g., the 9th cranial nerve is the glosso-pharyngeal, meaning nerve to the tongue and to the pharynx. Glossodynia means pain in the tongue.

*Vestibule*, 'the entrance', means that part of the mouth between the lips and the cheek on the outside and the teeth on the inside.

*Gingiva* (plural gingivae) is a term used to describe the tissue surrounding the teeth. It is roughly equivalent to the lay term 'gum'.

*Rugae* are transverse ridges of the mucous membrane which run across the anterior part of the hard palate.

*Deciduous teeth* are the first of the two sets of teeth that develop in man. They are also known as milk, baby, first and primary teeth.

*Permanent teeth* are the second of the two sets of teeth and are also called adult and secondary teeth.

The teeth are divided into four groups, incisors, canines, premolars and molars. The morphology of the teeth will be dealt with in Chapter 3 but it is worthwhile to note that there are considerable variations in size, shape and colour of the teeth between persons.

The *occlusal surfaces* of the teeth meet when the upper and lower jaws are brought together. *Occlusion* is the term for this bringing together of the jaws from a Latin word meaning 'to hide'. The occlusal surfaces are therefore hidden by occlusion. The front or anterior teeth have edges rather than surfaces and so it is more usual to speak of the *incisal edges* of the front teeth.

Each tooth has a *crown*, the part that is visible in the mouth when a tooth has erupted, and a root or roots which are implanted in the jaw bone. The junction between the crown and the root is known as the cervical or neck region and the end of the root is called the apex. At the apex there is a small opening or foramen through which blood vessels and nerves pass to supply the pulp or soft tissue within the tooth. The apical foramen is frequently divided up so that there are several small foramina.

A *cusp* is a conical elevation on a tooth surface, usually the occlusal surface. All teeth except the incisors have one or more cusps. The region between cusps often carries grooves of varying depth called *fissures*. These are important as often decay starts in a fissure. A *fossa* is used to describe a depression on the tooth surface which may have a pit in its depth. The *bifurcation* of the roots indicates the point at which the roots divide, where there are more than one root.

The *mesial* surface of the tooth is that surface of the tooth which is nearest the midline at the front of the jaws. Thus it is the medial surface of the anterior teeth and the anterior surface of the posterior teeth. The *distal* surface is that surface which is furthest from the midline in the jaws, i.e., the side opposite to the mesial surface (Fig. 1.1). The use of the word mesial is confusing as it is so near the word medial which the student will probably have learnt as a descriptive term in general anatomy. The use of medial in the present context would not be appropriate however in referring to the contact surfaces of the posterior teeth. These contacting surfaces are known as the *approximal* surfaces and the area between the teeth as the *approximal area*.

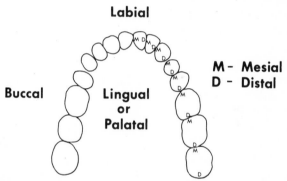

**Labial**

**Buccal**

**Lingual or Palatal**

**M -** **Mesial**
**D -** **Distal**

**Fig. 1.1** Diagram of the teeth to illustrate the descriptive terms of position

*Enamel* is the hard mineralised covering on the crown of a tooth. *Ameloblasts* are the cells responsible for enamel formation, ((en)ameloblast).

*Dentine* is the mineralised tissue which makes up the bulk of the tooth, covered in the crown by enamel and in the root by *cementum*. *Odontoblasts* are dentine forming cells. The *amelodentinal* junction (ADJ), the boundary between enamel and dentine, is an interesting junction as it is sensitive and pain is first felt when a dental surgeon drilling a tooth reaches it. The *pulp* is the soft tissue which is contained within the dentine of the tooth. It consists of nerves, blood vessels and several types of cells.

The *root canal* is the central canal of the root which is continuous with the pulp chamber above and with the soft tissues outside the tooth at the apical foramen or foramina.

*Cementum* is a mineralised, bone-like tissue which surrounds the root of the tooth and through which attachment to the jaw bone is effected. Sometimes it is loosely termed cement.

*Hydroxyapatite* is the main basic calcium phosphate crystalline material which is found in mineralising enamel, dentine and bone.

The *periodontal ligament* is the fibro cellular layer attaching the root of the tooth to the socket. It is also known as the periodontal membrane and may sometimes be seen as 'paradontal' in the American literature.

The *lamina dura* means 'hard sheet' and is a term used for the radioopaque line which is seen on radiographs around the root of a tooth corresponding to the bony boundary of a socket (Fig. 1.2).

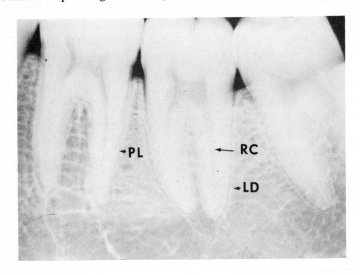

**Fig. 1.2** Radiograph of lower molars. LD=lamina dura; PL=periodontal ligament; RC=root canal

*Eruption* is the movement of the tooth towards or into the oral cavity. *Attrition* is a term used to describe the wearing down of teeth by other teeth. *Abrasion* is the wearing down of the tooth by an agent other than the teeth.

## Mastication

*Ingestion* is a term meaning the taking of food into the oral cavity. *Bolus* is the mass of food which has been masticated and is ready to be swallowed. *TMJ* is the *temporomandibular joint* between the mandible and the temporal bone. *Gnathostomatic system* is a sophisticated embracing term referring to all the chewing apparatus and includes teeth, muscles, bones, TMJ and all the nervous elements associated with it. *Articulation* is used in relation to the contacts that individual teeth make on opposing teeth. *Bite* is an old term meaning the position of the jaws when the teeth are in occlusion or in edentulous subjects the position the jaws would take up if teeth were present. This is an important position to establish when making dentures and is better described as an occlusal record. *Edentulous* is the state of being without teeth. The act of swallowing is called *deglutition*.

## The jaw bones

An *alveolus* is the socket for the tooth root and hence alveolar bone refers to the part of the jaw bone which carries the teeth. *Alveolar mucosa* is the mucosa which covers the jaw bone lying below the line of the attached gingiva. A *crypt* is the cavity or space within the bone in which resides a developing tooth. The term *ossification* means formation of bone and must not be confused with calcification which is the term for the laying down of calcium salts. The most commonly formed calcium salt is calcium phosphate which has several forms and which is found in bone, dentine, enamel and cementum, but calcification also occurs in some sites which do not become one of these recognised tissues, e.g., there may be calcification in cartilage or in ligaments or perhaps even in such sites as the kidney tubules in the form of renal calculi. The term *mineralisation* is roughly equivalent to calcification and is in fact more accurate in that other minerals in addition to calcium are often laid down during the process of 'calcification'.

## General

*Calculus* is a term used to describe tartar or hard formations that occur on teeth. *Plaque* is a soft material which accumulates on teeth and is mostly made up of bacteria. *Pellicle* is the thin film which forms on teeth almost immediately after brushing, and is present on all surfaces within the mouth. *Trauma* is the term used to describe any form of wounding of the tissues. *Lesion* is a term describing a wound caused either by trauma or by disease processes. *X-ray* is often used as a shorthand way of referring to a radiograph and in this case means the picture rather than the radiation.

It may be helpful for the student to know the names of disciplines within dentistry. *Conservation* deals with restoration of tooth substance which has been lost by *caries* (tooth decay) or by trauma. *Restorative Dentistry* is the discipline which deals with the replacement of teeth, which have been lost, by a prosthesis and hence is sometimes known as prosthetic dentistry. *Periodontology* is the study of diseases of the gums. *Orthodontics* is the study and treatment of abnormal occlusal relationships, i.e., the straightening of teeth which are malplaced or malpositioned. *Paedodontics* is dentistry for children. Other branches of dentistry such as Oral Surgery, Oral Medicine and Oral Pathology are self explanatory.

Other more specialised terms will be defined as they arise in the text but the student will find it useful to refer back to this chapter from time to time.

# 2

# The mouth

The mouth, or oral cavity, is bounded by the lips, cheeks, palate and the floor of the mouth. Posteriorly it is in continuity with the nasopharynx above and the oropharynx below.

Except for the teeth the mouth is lined by mucous membrane which is similar in its epithelial covering to that of skin. It differs however from skin in being moist, in its degree of keratinisation and its appendages.

### The lips
The junction between skin and mucous membrane is a sharp line known as the vermillion border. Here the keratin layer of the epithelium abruptly decreases in thickness so that the epithelium becomes more translucent and appears red. The lips are very sensitive and in infants they are the main exploratory areas before they learn to use their hands for this purpose.

The lip contains the orbicularis oris muscle and bundles of muscle fibres radiate from it to the bones round about. These muscles, especially the orbicularis, can exert considerable force, a characteristic of great importance from the orthodontists' viewpoint. By counteracting the thrust of the tongue the muscles help to maintain the anterior teeth in their proper relationships (Fig. 2.1). The muscles of the lips can sometimes be troublesome to the dental surgeon operating in the lower incisor region as they push the lip up and obscure his field of vision. The attachment of the radiating muscle fibres to the mandible and to the maxilla results in bands of mucosa-covered muscle crossing the vestibular sulcus in the midline and in the canine to premolar regions. In denture construction the base of the denture is cut back to permit free movement of the muscles in these areas otherwise dislodgement of the denture may occur with movement of the lips and cheeks.

The lips also contain on the oral aspect of the muscles, numerous minor salivary glands which can be felt by running a finger lightly over the surface. These glands often reflect disease processes in the major salivary glands and as they are easily accessible they are sometimes biopsied to help in the diagnosis of disease.

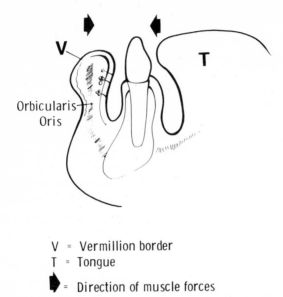

V = Vermillion border
T = Tongue
= Direction of muscle forces

**Fig. 2.1** Diagram of the forces on the anterior teeth. The position of the teeth is related to the balance between the thrust of the tongue and the pull of the lips

## The cheeks

The cheeks are similar in general structure to the lips but in the infant contain an appreciable amount of fat. This fat is said to be important in suckling as it lends substance and some rigidity to the side walls of the mouth. The cheek contains the buccinator muscle which acts in mastication to clear the vestibular sulcus and to squeeze food into the space between the occluding teeth. Teeth may alter position in the bone if forces are applied to them. The bucco-lingual position of the cheek teeth is partly determined by the balance of forces between the tongue on one side and the buccinator muscle on the other, just as the balance between the tongue and the lips governs the position of the anterior teeth. The mucous membrane of the cheek is similar to that of the lips having a covering of nonkeratinising epithelium and minor salivary glands with numerous ducts piercing the lamina propria. The surface layers of the epithelium desquamate into the saliva and cells from the surface can be readily collected by rubbing a finger or a spatula across the surface and smearing onto a microscope slide. The study of cells by this technique is known as exfoliative cytology. It has the advantage of not creating a wound nor does the technique require any anaesthetic. Normally the replacement of these desquamated cells occurs by proliferation in the basal layer of the epithelium which in the mouth generally has a higher turnover rate than in the epidermis.

### The gingivae

The oral mucosa covering the alveolar bone, also known as gum or gingivae, is tightly bound to the underlying bone. It is pink and has a stippled appearance in health. Round the necks of the teeth the gingiva forms a shallow groove about 1–2 mm in depth with the tooth as the inner wall. This is the gingival sulcus or crevice and deepening of this sulcus producing a pocket is an early stage in the development of periodontal disease. Measurement of the depth is therefore an important procedure in the examination of a patient. On the oral surface of the gum a linear shallow depression following the line of the crest of the gingiva and approximately 2 mm below it, indicates the limit of the 'free gingiva'. The gingiva more apically is attached firmly to the bone and hence is called attached gingiva. The attached gingiva ends abruptly at the mucogingival line by meeting the alveolar mucosa, a much looser tissue which is darker in colour and which is continuous across the vestibular sulcus with the labial or buccal mucosa (Fig. 2.2).

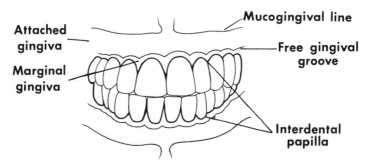

Fig. 2.2 Diagram of the anterior teeth and gums

### The palate

On the palatal side of the upper teeth the attached gingiva is continuous with the mucoperiosteum covering the hard palate. In the incisor region the mucosa is tightly bound to the bone and injections to anaesthetise the nerves which enter the palate from the incisive foramen, can be painful. Posterior to the first premolar tooth there is a looser tissue at the sides of the palate, and injection of local anaesthetic solution here does not cause quite so much pain (Fig. 2.3).

A variable number of ridges run transversely across the palate.

They are known as rugae and have a core of dense connective tissue. There is considerable individual variation in the pattern of rugae and they have been called the finger prints of the mouth. In the infant they provide a rough surface to grasp the nipple and in the adult they assist in mastication providing a rough surface against which the tongue can press the food. They are highly developed in many animals and are richly provided with tactile sensory organs. They may also contain taste buds.

### Section through the palate

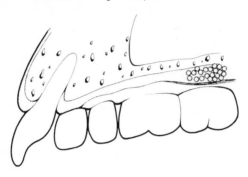

**Fig. 2.3** Parasagittal section through the upper canines. Anteriorly the palatal mucosa is tightly bound down. Posteriorly glands and looser tissue separate the mucosa from the bone

Behind the hard palate and continuous with it is the soft palate, which has a core of muscles and minor salivary glands. The soft palate has a free posterior border elongated centrally to form the uvula. It can be raised or lowered during speech to create a variation of the sound emitted from the larynx. During swallowing it is raised to prevent solid and liquid foods from the entering the nose. These movements are made possible by the levator palati muscle which is inserted into the lateral part of its upper surface, the palato-glossus muscle which runs from its undersurface down laterally to the tongue and the palato-pharyngeous muscle which runs downwards from the palate to the side walls of the pharynx. The core of the soft palate is strengthened by the tendon of the tensor palati muscle which uses the lower end of the medial ptyerygoid plate (the hamulus) as a pulley to enter the soft palate from the side. To complete the musculature picture a rather weak muscle runs from front to back in the central area of the soft palate, the uvular muscle. Apart from tensor palati which is supplied by the trigeminal nerve the muscles are innervated by the vagus and cranial accessory nerves. The tensor palati is attached to the lower part of the Eustacian tube and when it contracts this tube is made patent. Thus the pressure of air in the pharynx and the middle ear is equilibrated. Thus swallowing while the air pressure in the environment is changing restores the ability of the tympanic membrane to respond to sounds. This explains why discomfort in the ears produced when climbing a mountain or flying may be relieved by swallowing.

The two muscles which run from the undersurface of the soft palate, the palato-glossus and the palato-pharyngeus raise vertical folds of mucous membrane on the side walls of the posterior part of the oral cavity. On each side between the folds, which converge above, is the tonsil, a mass of lymphoid tissue lying under the epithelium. This tonsil is part of a ring of tonsillar or lymphoid tissue encircling the opening into the oropharynx. The ring is completed by the adenoids a similar mass of lymphoid tissue on the posterior wall of the pharynx, and the lingual tonsil lying under the epithelium on the back of the tongue. The function of the tonsillar ring is not completely understood. Medical opinion is divided on the need for removal of the tonsils when infection occurs during childhood. The masses of lymphoid tissue tend to regress in middle and old age, but it is not uncommon for young adults to suffer from tonsillitis with tonsillar enlargement and consequent restriction of the width of the oropharynx.

**The tongue**
The tongue is a strong muscular organ which almost fills the space bounded by the teeth when the mouth is closed. The surface at the sides is often marked by the teeth in contact with it at rest. Over the dorsal or upper surface the epithelium bears two types of papillae known as filiform (wire like) and fungiform (mushroom shaped) respectively (Fig. 2.4). The latter appear as small red dots which are very numerous near the tip and along the sides. Taste buds are associated with these papillae. The filiform papillae are much more numerous and have a greyish colour due to the cap of keratin which surmounts them. They give the tongue its rough surface and are much more developed in some animals, for example the cat, which uses its rough surfaced tongue to lap up milk. In the human, illnesses which affect the alimentary canal cause the papillae to become long and coated with bacteria. An experienced medical practitioner can gain a lot of information from the condition of his patient's tongue.

Dividing off the anterior two thirds from the posterior third of the tongue is a V shaped line of small circular structures similar in shape, but larger than the fungiform papillae. These are known as circumvallate papillae and they also carry taste buds. They are described in more detail in Chapter 7. Behind this the tongue has a pebble stone

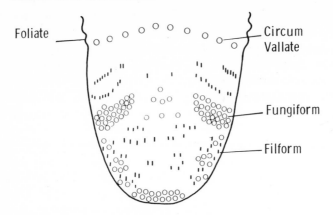

Distribution of Papillae on tongue

**Fig. 2.4** Dorsal surface of the tongue. The fungiform papillae in life appear as red spots and are much less numerous than the filiform papillae

appearance due to the lymphoid tissue underlying the epithelium, the lingual tonsil.

Posteriorly the side walls of the tongue sometimes have vertical folds of mucous membrane which are called foliate (leaf shaped) papillae. This is an area which can be irritated by rough or sharp edges on the adjoining teeth. It is commonly inflammed and can give rise to considerable discomfort. It is important that any dental examination should include an inspection of this area of the tongue as it is the most common lingual site for carcinoma.

The undersurface of the tongue has a thinner epithelial covering and through the epithelium can be seen large veins. In the midline a thin vertical fold of mucous membrane from the tongue is attached to the inner aspect of the mandible. This is known as the frenum. The muscles of the tongue either come from structures outside it, the extrinsic muscles, or begin and end within it, the intrinsic muscles. All these muscles enable the tongue to carry out very complicated movements and also to change its shape, an important characteristic in the production of intelligible speech. The extrinsic muscles approach the tongue from in front, behind, below and above, thus ensuring a high degree of mobility. In front the genioglossus runs from the mandible and by its action pulls the tongue forward. The styloglossus muscle enters the tongue from its attachment to the styloid process and can pull the tongue backwards. The hyoglossus muscle arises from the hyoid bone and pulls the tongue downwards. Superiorly the palatoglossus muscle is attached to the palate and it raises the posterior part of the tongue. The intrinsic muscles are orientated in three planes namely, vertical, transverse and longitudinal and are attached to the mucosa and a midline fibrous septum. The first three extrinsic muscles together with all the intrinsic muscles develop from myotomes which have migrated from the occipital region in the embryo and hence are innervated by the hypoglossal or twelfth cranial nerve.

Scattered among the muscle bundles are large numbers of small mucous glands which open by numerous small ducts onto the surface of the tongue. A special group of serous glands which open into the circular troughs around the circumvallate papillae at the back are known as Von Ebner's glands. They serve to wash out these troughs where the taste buds are located, presumably to prevent the sapid substances from stagnating in the relatively deep troughs.

The tongue in mastication presses the bolus of food against the hard palate and assists in keeping the food between the teeth during chewing. Its mobility is also used in clearing the vestibular sulcus of food debris. The rich sensory innervation enables information on the consistency of the food during mastication to be sent to the centres controlling reflex mastication. Proprioceptive innervation of the musculature of the tongue is of great importance in correct position-ing of the tongue during the complicated movements occurring during speech. It is interesting that the ability to perform certain types of movement of the tongue is genetically inherited. The tongue is ex-tremely vascular and bleeds readily when cut. The main arteries are the right and left lingual arteries, branches of the external carotids. They run nearer the undersurface than the dorsal surface parallel to but deeper than the lingual veins. The lymphatic drainage of the tongue is important in view of its predilection as a site of cancer. The tip of the tongue has lymphatics which drain into the submental glands and from the sides they drain to the submandibular glands. Some lymphatics from the tip may run directly to the cervical glands.

**Floor of the mouth**
The floor of the mouth is covered by a thin smooth epithelium through which can be seen the underlying blood vessels giving it a dark red appearance. The sublingual and part of the submandibular salivary glands lie below the mucosa. Their ducts open in little papillae on either side of the frenum of the tongue behind the incisor teeth (Fig. 2.5). Stretching posteriorly and laterally from this area are folds of mucous membrane, the fimbria. The salivary glands lie on the mylohyoid muscle which is attached to the inner surface of the mandible and the upper surface of the hyoid bone. This muscle aids in depressing the mandible if the hyoid bone is fixed by the infra-hyoid

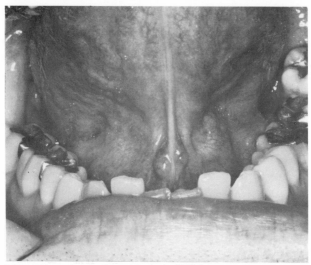

**Fig. 2.5** Floor of mouth. The lingual frenum separates, at its lower end, the two openings for the submandibular ducts

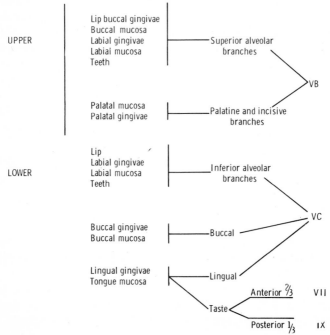

**Fig. 2.6** Scheme for the sensory innervation of the oral mucosa. VB=Maxillary division of the trigeminal nerve; VC=Mandibular division; VII=Facial nerve; IX=Glossopharyngeal nerve

muscles. In swallowing it assists in raising the hyoid and with it the larynx when the jaws are closed.

The sensory innervation of the mouth is mainly from the fifth cranial nerve. Contributions from the seventh in the form of special taste sensation and from the ninth as common and special sense complete the sensory picture (Fig. 2.6).

The anterior part of the mouth is the most sensitive, the upper lip being well supplied with sensory nerves especially along the vermillion border. It has been shown by the two point discrimination test that the degree of sensitivity progressively decreases towards the back of the mouth. This two point test conisists of applying two movable points to the area to be tested and by gradually moving them closer, the minimum distance for discriminating two points rather than one can be determined. This minimum distance is considered to be inversely related to the degree of sensitivity.

The mouth has a higher pain threshold than the skin for hot and cold stimuli. It is possible to drink fluids at a higher temperature than can be borne by hand. The reason for this is not known but it has been suggested that there is a higher degree of vascularity in the mouth and this ensures rapid dispersion of heat or warming of cold objects. Probably saliva plays a part also in cooling or warming the incoming fluid when the temperature tends to deviate from normal.

The teeth and surrounding gums are dealt with in later chapters.

# Appendix: Local anaesthesia

The dental surgeon uses local anaesthesia every day in his practice and thus it is important to know the nerve supply to the various structures he might want to anaesthetise. Local anaesthetics block impluses along nerves and usually contain a vaso-constrictor so that the blood supply locally is reduced. This has the effect of lengthening the duration of anaesthesia. The sensory nerve supply to the mouth is almost entirely from the mandibular and maxillary divisions of the trigeminal nerve.

The mandibular nerve supplies the gingivae and teeth in the lower jaw through the long buccal nerve, the inferior dental nerve and the lingual nerve. The mandibular bone is dense on its buccal and lingual aspects and anaesthetic solutions do not penetrate easily. The inferior dental nerve is usually blocked for local anaesthesia before it enters the mandible at the inferior dental foramen. The lingual nerve which supplies the gum on the lingual side of the mandible in addition to the tongue, can be anaesthetised here also, as it lies close to the inferior dental nerve (Fig. 6.1). Note that the lingual nerve has an important relationship to the mandible on the lingual side of the alveolar bone surrounding the wisdom tooth. Care must be taken not to damage it whilst extracting this tooth. The mental foramen transmits the mental nerve which supplies the gum on the labial side of the anterior teeth and the lower lip. It is possible to anaesthetise the teeth in front of the mental foramen by injections into this foramen. In the incisor region the mandibular cortex is thinner and deposition of anaethetic solution outside the bone can often produce anaesthesia of these teeth. This is known as infiltration anaesthesia as opposed to 'block' anaesthesia of the nerve before it enters the bone.

In the upper jaw the nerves to the teeth come from the anterior middle and posterior dental branches of the maxillary division. The maxilla does not have such a dense cortical bone as the mandible and infiltration anaesthesia in the region above the apex of the tooth is usually sufficient. On the palatal side the gingivae from the canine backwards are supplied by the greater palatine nerve, which runs parallel to the alveolar margin and approximately midway between the margin and the midline. Anaesthetic solutions are best placed posteriorly as there is less pain with injections in that region. The palatal gingivae of the incisors are supplied by the incisive nerve, a terminal branch of the long sphenopalatine nerve. Injections in this area are painful due to the tight binding of mucosa to periosteum. Infiltration anaesthesia on the buccal side is sufficient for conservation of the teeth but for extractions and palatal surgery the palatal nerves must also be anaesthetised.

# 3

# The teeth

Teeth are generally considered to have evolved from epidermal scales which were situated around the mouth in our aquatic ancestors. Many primitive animals have a homodont dentition, that is all the teeth are similar to each other in form. With time animals evolved with a heterodont dentition, where teeth vary in form in different parts of the mouth. In man the teeth are less specialised than in many other animals. There is some degree of variation in that the anterior teeth are adapted for incising and the cheek teeth are adapted for crushing and mixing. Another change is that man in common with most mammals now has only two sets of teeth, deciduous and permanent, in place of the endless succession found in primitive animals. The first dentition lasts for a comparatively short time and holds the fort as it were until the jaws are large enough to accommodate the permanent teeth.

There are thirty-two teeth in the permanent dentition and these are symmetrically arranged with eight teeth in each of four quadrants, right and left, upper and lower (Fig. 3.1). This division into four quadrants is the basis of the most common form of notation for the teeth used in clinical dental practice in Great Britain. The patient's

$$\frac{8\ 7\ 6\ 5\ 4\ 3\ 2\ 1\ \mid\ 1\ 2\ 3\ 4\ 5\ 6\ 7\ 8}{8\ 7\ 6\ 5\ 4\ 3\ 2\ 1\ \mid\ 1\ 2\ 3\ 4\ 5\ 6\ 7\ 8}$$

**Fig. 3.1** The horizontal line separates upper from lower and the vertical line represents the mid line

right is on the left of the diagram since the teeth are represented as viewed by the dental surgeon from in front of the patient. The tooth nearest the midline is known as the central incisor sometimes abbreviated to 'central'. The next tooth in the arch is known as the lateral incisor or more simply, 'lateral'. The third tooth from the midline is the canine or in North American terminology the cuspid.

Behind this tooth are the two premolars or bicuspids followed by the molars of which there are three.

A shorthand way of referring to a single tooth is to use only two bars of the cross, e.g., 1/ is the symbol for the upper right central incisor.

The deciduous dentition contains twenty teeth as there are no premolars and only two molars in each quadrant. For the deciduous teeth letters are used instead of numbers to distinguish them from the permanent dentition but with the same framework denoting each side and each jaw (Fig. 3.2). Thus the lower right central incisor in the deciduous dentition is designated as A̅/. The student should practise the use of this notation until he is familiar with it.

$$\begin{array}{ccccc|ccccc} E & D & C & B & A & A & B & C & D & E \\ \hline E & D & C & B & A & A & B & C & D & E \end{array}$$

Fig. 3.2

## Morphology of the individual teeth

As mentioned above the teeth can be subdivided into four groups, incisors, canines, premolars and molars. Individual members of these groups have general characteristics of form by which they can be recognised. There is considerable variation in form within the general characteristics of the type however and this can give rise to problems in tooth identification. A knowledge of tooth morphology is needed when the dental surgeon restores a broken down tooth or when he has to prescribe treatment to be undertaken by a colleague. The dental surgeon uses this knowledge when he distinguishes between deciduous and permanent teeth and will find the knowledge useful if he is involved in forensic work.

*The incisors* (Figs. 3.3a, b)
The incisors are single rooted teeth having the crown shaped like a chisel at the incisal edge. The upper central incisor crown when viewed from in front has an approximately rectangular form. The mesial margin and the incisal edge are fairly straight whereas the distal margin is slightly convex and the upper or cervical margin is markedly convex. The corner where the mesial margin and incisal edge meet is almost a right angle in contrast to the distoincisal corner which is more rounded. The difference between the incisal corners helps to distinguish left from right upper central incisors. In some cases there

are two shallow grooves running from the incisal edge down the labial surface partly dividing it into three parts.

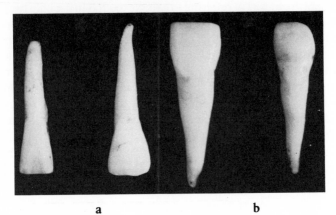

**a**                              **b**

**Fig. 3.3a** Upper right incisors. The distal corner of the incisal edge is rounded

**Fig. 3.3b** Lower right incisors. The lateral is larger than the central

The palatal surface is triangular in form with the incisal edge forming the base and the cingulum forming the apex of the triangle. The cingulum is a pronounced bulge of enamel near the cervical margin and from it mesial and distal marginal ridges run to the incisal edge bounding a central concave area. Occasionally there is a small pit in the enamel in this concave area where the marginal ridges converge, forming a site where decay may start. Exaggeration of the marginal ridges occurs in teeth of Mongoloid races producing so-called 'shovel-shaped' incisors.

The mesial and distal surfaces of the crown are triangular in shape and incline palatally. They are shorter than the labial and palatal surfaces as the crown root junction follows a sinuous course around the tooth.

The gum is attached to the tooth, in life, close to this crown root junction. Because of the sinuous nature of the junction there are high 'inter-dental papillae' between teeth and a scalloped gum margin on both labial and palatal surfaces. The sinuous pattern of the gum margin is found to some extent on all teeth but it is most marked on the incisors and canines. Knowledge of its position is of aesthetic importance in designing restorations of the upper anterior teeth and in making dentures to replace them.

The root of the tooth is stout and round in cross section tapering gradually to the apex or tip of the root. The pulp of the tooth is a box like cavity in the crown portion and is continuous with the central root canal which is circular in cross section. From the roof of the pulp in the crown portion small extensions reach out towards the mesial and distal incisal corners. These are known as the cornua of the pulp and care must be taken to avoid them when preparing the teeth for restorations.

The upper lateral incisor tooth is similar in form to the central incisor except that it is narrower mesio-distally. Sometimes it is missing completely from the dentition. The crown is more variable in its shape than the central incisor and may even have a simple peg shape. The distal corner of the incisal edge is more markedly rounded than that of the central incisor and the contrast between this rounded corner and the sharp mesial corner again makes the determination of left and right sides easy. As with the central incisor the palatal surface often has a small pit where the marginal ridges meet the cingulum. This pit occasionally is very deep and can lead to pulp death.

The pulp cavity tends to be large relative to the size of the tooth. The root of the tooth is almost circular in cross section and may be almost of the same length as the root of the central incisor. The root inclines palatally in the patient and an abscess which forms on this tooth is liable to appear as a swelling in the roof of the mouth.

The lower central incisor is the smallest tooth in the permanent dentition. The crown is long and narrow with the incisal edge meeting the mesial and distal surfaces almost at right angles. The distal incisal angle may be slightly more rounded than the mesial. There is little or no cingulum and marginal ridges are not well developed on the lingual surface of the tooth. The pulp chamber has two cornua directed towards the mesial and distal angles of the incisive edge. Whereas the crown of the tooth is flattened bucco-lingually the single root is flattened in the mesio-distal direction and is oval in cross section.

In the lower jaw the lateral incisor is wider than the central incisor, contrary to the situation in the upper jaw. The mesial and distal sides of the crown tend to diverge incisally giving a fan shaped appearance to the crown. The distal corner of the incisal edge is more rounded than the mesial corner, and the crown appears to have been rotated slightly on the root. It is often difficult to distinguish between the lateral and central lower incisors and between right and left teeth. The root of the lateral is similar to that of the central.

All of the incisor teeth when first erupted may bear three small elevations or cusps on the incisal edge. These small cusps are known as mammelons and are soon worn away with attrition.

## The canines (Fig. 3.4a, b)

The canines or cuspids are the longest teeth in the permanent dentition. They are named from the very prominent corresponding teeth in dogs (canines). Carnivorous animals with long canines are restricted in the amount of lateral movement of the mandible but in man the canine is at the level of the occlusal plane and so a greater range of movement of the mandible is possible. In physical anthropology the level of the canine above the height of the other teeth helps to distinguish ape-like from hominid skulls.

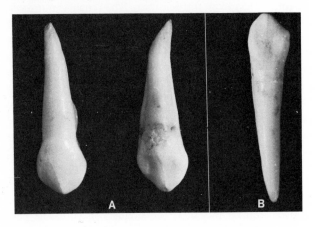

**Fig. 3.4a** Upper canine labial and palatal views
**Fig. 3.4b** Lower canine lingual view. There is no cingulum on this tooth

The upper canines are sometimes referred to as the 'eye' teeth but contrary to popular belief they do not correspond to the piercing teeth of the vampire bat which are in fact lateral incisors. The crown has a high pointed cusp with mesial and distal incisal ridges. The distal incisal ridge is the longer of the two. The labial surface has two shallow vertical grooves with a low spine between them running from the cusp towards the cervical margin. The distal border of the labial surface is markedly convex and hence it is relatively easy to tell from which side the tooth comes. Palatally the crown bears a well marked cingulum which adds to the bulk and strength of this tooth. From the cingulum a central ridge runs to the apex of the cusp and this ridge together with the mesial and distal marginal ridges create two fossae on the palatal surface. Overall the crown has a short 'fat' appearance. The pulp cavity is large, tapers towards the summit of the crown but does not have cornua as in the incisor teeth. The single root is large and strong. It is almost triangular in cross section, the apex of the triangle facing palatally.

The lower canine has a longer and narrower crown compared with the upper, and the lingual surface is devoid of a cingulum (Fig. 3.4b). The distal surface is more convex than the mesial surface and the cusp tip lies nearer the mesial side of the tooth. An important clue to the identity of this tooth can be found in the position of the 'wear facet' on the crown. In the mouth the upper canine crown overlaps the labial surface of the lower and is slightly behind the lower so that a flat area is worn on the distal part of the buccal surface of the lower. A corresponding wear facet is usually found on the palatal surface of the upper canine on the mesial slope of the cusp. Thus, if the wear is on the buccal side the canine is probably from the lower jaw and vice versa. The convexity of the crown root junction is more marked on the mesial than on the distal surface. The pulp chamber is narrower than in the upper canine but has a similar taper towards the summit of the crown. The root is flattened on the mesial and distal surfaces which are sometimes grooved longitudinally. Occasionally the root may be bifid, i.e., the grooving on mesial and distal surfaces may be so deep as to result in two separate roots.

The canines and incisors are sometimes grouped together as the anterior teeth and are obviously of great aesthetic consideration especially the uppers. The upper lip has more mobility than the lower lip and hence these teeth show more than the lowers. The colour of the incisors is not uniform, the incisal edge often having a faintly darker hue and the tooth being slightly yellower towards the neck. These nuances of colour are difficult to reproduce in artificial teeth hence the latter are fairly easy to detect unless great care is taken in their manufacture. The canine teeth are often slightly more yellow than the incisors and this should be borne in mind in trying to induce a natural appearance in a denture and in reassuring patients who worry about the canines being too yellow.

The shape and size of the teeth are genetically determined and hence have a relationship to the physical build of the individual. Choosing the correct mould of artificial teeth for a patient where the natural teeth are not available to act as a guide, is an art rather than an exact science. In some texts it is suggested that the shape of the patient's head inverted should be the shape of the central incisor, but personal observation of the varied natural dentitions in the people he meets will be of greater help to the student than any specified set of rules.

*The premolars* (Figs. 3.5, 3.6)
The premolars or bicuspids have typically two cusps placed buccally and lingually. In the lower jaw there is a gradual development of the lingual half of the cheek teeth from a small lingual cusp on the first premolar to a full double cusp on the lingual half of the molar teeth. This trend towards an occlusal surface starts earlier in the upper jaw

with the cingulum on the upper canine and the high palatal cusp of the first premolar. The crowns of the upper premolars sit squarely on the root so that the midline on the mesial or distal surface of the tooth passes through the fissure between the cusps. The lower crowns are tilted lingually so that the buccal cusps lie over the midline of the root rather like the student who has fallen asleep at the lecture. The cervical margins of these teeth are less sinuous than in the incisors or canines so that the scalloping of the gingivae is shallower than in the anterior region.

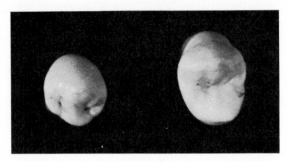

**Fig. 3.5** First premolars. The lower premolar has a more circular crown and the occlusal surface is dominated by the buccal cusp

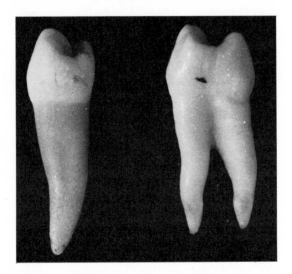

**Fig. 3.6** A comparison of upper and lower premolars. The lower premolar with its single root has the buccal cusp over the midline of the root. The upper premolar has the two cusps evenly spaced from the midline of the tooth

The first upper premolar is a rugged tooth having a large buccal and a smaller palatal cusp. The apex of the buccal cusp is nearer the distal than the mesial surface so that the mesial slope of the cusp is longer than the distal. Between the two cusps the central fissure ends in a small pit before the distal marginal ridge but mesially it spills over the marginal ridge before ending on the mesial surface. This mesial surface shows a distinct shallow concavity near the cervical margin which is known as the 'canine fossa'. This fossa gets its name from its position and not because it is produced by the distal bulge of the canine. The distal surface of the canine in fact contacts the first premolar above this concavity. The fossa makes it difficult to restore this surface of the tooth without leaving an overhang at the cervical margin. The pulp chamber is elongated bucco-palatally with cornua directed towards the cusps. The root portion has distinct vertical grooves on the mesial and distal surfaces and is often divided into two separate roots. When two roots are present the distance of bifurcation from the cervical margin of the crown may vary. Two root canals are common even with single rooted teeth.

The upper second premolar has a similar crown pattern to that of the first. If the occlusal surface of the first premolar is considered rugged and 'masculine' the second premolar has a more rounded or ovoid 'feminine' appearance. The buccal and palatal cusps are more nearly equal in size than in the first premolar and the tips of both cusps tend to be situated nearer the mesial than the distal surface. There is a small central fissure on the occlusal surface which has distinct mesial and distal marginal ridges. There is no concavity on the mesial surface similar to that on the mesial surface of the first premolar. The pulp chamber is similar to that in the first premolar except that there may be only one root canal. The root is nearly always single with vertical grooves on the flattened mesial and distal surfaces.

The lower first premolar has a high buccal cusp and a much lower lingual cusp which resembles an exaggerated cingulum. The two cusps are connected by a central ridge which divides the occlusal surface into a small mesial part and a large distal part. From the occlusal surface a fissure runs over the mesial marginal ridge onto the mesial surface which faces lingually as well as mesially. As mentioned earlier the crown is tilted lingually on the long axis of the tooth and the buccal surface of the crown and root form a continuous convex curve when viewed from the mesial or distal side. The pulp chamber has a cornu extending towards the buccal cusp and there is usually a single root canal. The single root is flattened mesio-distally with shallow vertical grooves on the mesial and distal surfaces.

The second lower premolar has two cusps which are nearly equal in size, though the lingual inclination of the crown on the root enhances the appearance of a larger buccal cusp. The occlusal surface is sometimes divided by a central ridge of enamel, producing a large distal and a small mesial fossa. The lingual cusp is closer to the mesial than the distal surface and sometimes is subdivided by an extension of the central fissure into a larger mesial part and a smaller distal part. The pulp chamber is oval in cross section with cornua directed towards the cusps. The root is similar to that of the lower first premolar and has a single or partially divided root canal.

## The molars (Figs. 3.7, 3.8)
The molars or grinding teeth are found behind the premolars. As the name suggests they grind food between occlusal surfaces which bear between three to five cusps. In the upper jaw the first molar is the largest and the third the smallest of the three (Fig. 3.7). In the lower jaw the first is usually larger than the second but the third is very variable in it size and shape. The upper molars usually have three roots and the lower molars two. The crowns of the upper molar teeth have an anterior triangular element of three cusps, two buccal and one palatal. This triangular pattern is found in many other primates and is thought to be the basic pattern in the evolution of cheek teeth. The crowns of the lower molars tend to be more rectangular with a central fissure running mesio-distally.

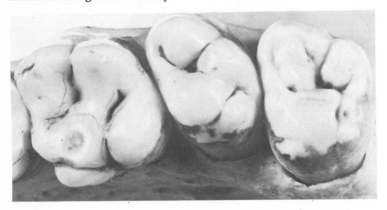

**Fig. 3.7** Photograph of maxillary molars in position in a skull. The oblique ridge is well shown in the first and second molar. The third molar is almost triangular in occlusal shape

The upper first molar has an occlusal surface divided into an anterior triangular part and a smaller posterior part known as the talon or the heel. A ridge from the disto-buccal cusp to the mesio-palatal cusp marks this division of the occlusal surface. Two fissure systems can thus be found. A mesial fissure runs from the central fossa onto the buccal surface between the two buccal cusps where it sometimes ends in a small pit. The distal fissure runs from the small distal fossa to the palatal surface. These fissure systems are commonly the site of carious attack. The mesio-palatal cusp is the largest of all four cusps. On its palatal slope is sometimes found a secondary enamel elevation which can be any size from a small outcrop of enamel to a cusp almost as high as the mesio-palatal cusp itself. This secondary cusp rejoices in the name of 'Carabelli's tubercle' or 'cusp of Carabelli'. The pulp chamber of this tooth is almost cubical in form with cornua projecting towards the cusps. From the floor of the pulp chamber there are openings into the three root canals. The three roots arise from the broad area of the root region at a short distance below the crown. The two buccal roots, placed mesially and distally, are flattened in the mesio distal diameter and the mesial root is usually the longer of the two. The palatal root is more rounded and diverges markedly from the other two. It is important to note that the distance between the buccal and palatal roots at their apices is greater than the diameter of the crown at the neck, an important consideration in extracting this tooth.

The upper second molar has a crown form which is very similar to that of the first molar. There is a tendency for the occlusal surface to be more markedly rhomboidal, with the mesio-palatal to disto-buccal diameter becoming shorter. The disto-buccal and disto-palatal cusps are usually reduced in size compared to the first molar. The pulp chamber is similar in form to that of the first molar with cornua again extending towards the cusps. The roots of the tooth are less divergent than those of the first molar. Sometimes the palatal and a buccal root are fused together though usually there are still three root canals.

The upper third molar shows great variation in its crown form. A triangular pattern with one palatal cusp and two buccal cusps is the commonest type with the mesial being the larger of the two buccal cusps. Very occasionally a cusp of Carabelli is present on this tooth. The three roots are rarely divergent and are most often found fused together with a distinct distal curve near the apex.

The occlusal surface of the first lower molar (Fig. 3.8) has a rectangular outline with five cusps, three placed buccally, and two which are slightly more pointed, placed lingually. The cusps decrease in size from before backwards and the distal of the three buccal cusps is almost mid-way between the buccal and lingual surfaces. There are distinct mesial and distal marginal ridges. The occlusal surface is divided by a central fissure running mesio-distally with branches running between the cusps onto the buccal and lingual surfaces. A small pit which often becomes decayed may be present on the buccal surface at the end of the fissure. There is a slight lingual inclination of the crown on the root and thus the buccal surface of the crown inclines lingually whereas the lingual surface is almost vertical. The pulp chamber is box shaped with cornua projecting towards the four main cusps. There are two stout roots placed mesially and distally both being oval in cross section with the long axis of the oval lying bucco-lingually. The mesial root usually curves distally and the distal root is straight. Both roots are grooved vertically on the mesial and distal surfaces. There are usually two root canals mesially and one distally. The distal canal is oval in cross section.

**Fig. 3.8** Lower molars *in situ*. The first and third have five cusps each, the second has four cusps

The lower second molar has also a rectangular occlusal surface but is smaller than the first molar. It has four cusps with a cruciate fissure pattern between them. The mesial cusps are the larger pair and the buccal surface is more convex from top to bottom than the lingual surface. A side arm of the central fissure runs out to the buccal side between the buccal cusps just as in the first molar. The roots are similar to those of the first molar but are shorter and more likely to be fused together.

The third molar or 'wisdom tooth' is very variable in its form. Probably the commonest type is where the crown resembles the first molar with five cusps. A circular crown is sometimes seen with a ring of small cusps around the central pit. The distal surface is usually markedly convex, bucco-lingually and from top to bottom. Occasionally the tooth is larger than the second molar. The roots also show great variation. At one extreme they are well formed and may resemble those of the first molar in size and shape. At the other end of the spectrum and more usually they may be fused together into a small conical mass. Since the jaws of man have become smaller in evolution and the teeth have lagged behind in the reduction of size there is often not enough room in the mandible for the third molar to erupt. The tooth is then said to be impacted, that is, it is in contact with the tooth in front or the bone behind in such a way that its eruption path is impeded. In these cases the morphology of the crown and more especially the shape and number of roots are prime factors in determining the difficulty of extraction. Extraction is further complicated by the relationship of the roots of the third molar to the inferior dental nerve. At times they are grooved by the nerve and may even be perforated by it.

### The deciduous dentition

The deciduous teeth have several characteristics which enable the student to distinguish them from the permanent teeth. In colour they are whiter than the permanent teeth because the enamel is more opaque. The crowns of the deciduous teeth are more bulbous than those of the permanent teeth and at the neck of the tooth the enamel meets the cementum at a greater angle than in the permanent teeth (Fig. 3.9). This junction is not as sinuous as in the permanent teeth.

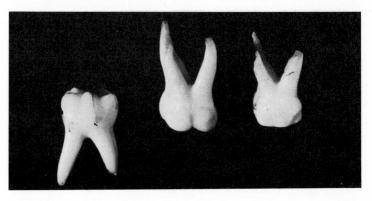

**Fig. 3.9** Deciduous molars. From left to right lower second, upper second, upper first. The wide divergence of the roots and the cervical constriction of the crown are typical of these teeth

The cusps are sharper in the deciduous dentition when they first erupt but they are worn down more quickly than the cusps of the permanent teeth as the enamel is softer. The deciduous incisors and canines are smaller than their successors and the deciduous molars are smaller than the permanent molars. Related to the size of the crown the deciduous roots are longer than the permanent roots but they are slender and have often undergone resorption by the time the teeth are shed.

The pulp chambers of the deciduous teeth are all much larger relative to the size of the crown than is the case with the permanent teeth, an important clinical consideration. As well as being less hard the enamel of deciduous teeth is more permeable than that of permanent teeth.

### Deciduous incisors

The labial surface of the upper central incisor is almost square and the distal corner of the incisal edge is quite rounded. It has a root form which resembles that of the permanent tooth. The lateral incisor is similar in shape to the permanent tooth except that the breadth of the labial surface is relatively greater. The lower incisors are also very similar to the permanent teeth except for their smaller size and the bulbous crowns.

### Deciduous canines

The upper deciduous canine is a squat tooth with a pointed cusp when first erupted. There is a marked cingulum. The lower canine is a longer tooth than the upper but is more squat than the permanent successor. The deciduous canines have relatively long roots. Deciduous anterior teeth are scarce as they are often extracted on a do-it-yourself basis and sold to the fairies.

### Deciduous molars

The upper first deciduous molar is unlike any of the permanent teeth. It has on its occlusal surface two cusps placed buccally and palatally with a broad shallow central fossa. The buccal side is longer than the palatal side and the mesio-palatal angle is obtuse. The buccal cusp which might almost be called a ridge is sometimes subdivided into two or three lobes. There is a pronounced bulge of enamel at the mesio-buccal corner of the crown near the cervical margin, known as the tubercle of Zuckerkandl. There are three roots, mesio-buccal, disto-buccal and palatal which diverge widely almost as soon as they arise from the cervical margin of the crown, and enclose in life the developing crown of the first premolar tooth.

The lower first deciduous molar has an occlusal surface which is longer mesio-distally than bucco-lingually. Two cusps are found on the lingual aspect of the occlusal surface, and two cusps on the buccal aspect though these may not be well developed. The mesio-lingual cusp is the largest. The occlusal surface is sub-divided into a small mesial and a large distal fossa by a ridge running from the mesio-lingual to the the mesio-buccal cusp. The buccal surface of the crown of this tooth has a very pronounced bulge near the cervical margin at the junction between the mesial surface and the buccal surface. This is similar to the bulge on the upper tooth but is relatively more extensive extending down the root. The two roots placed mesially and distally diverge and leave room for the crown of the developing first premolar. The divergence starts at the base of the crown with little or no undivided root portion.

The second deciduous molars both upper and lower closely resemble the crown patterns of the first permanent molars in the respective jaws except for the general differences stated earlier. In the second upper deciduous molar there may be a tubercle or cusp of Carabelli present. The upper molars have three roots widely divergent at first and sometimes turning in at the apex like the grab of a crane. The divergence starts immediately below the crown and allows room for the developing crown of the second premolar which succeeds the molar tooth. The lower molars have two roots placed mesially and distally and in these teeth also the roots diverge to surround the developing second premolar crown. All of the deciduous roots undergo some degree of resorption before they are shed.

The size of the pulp chamber in deciduous teeth has already been mentioned but it is worth repeating now in view of the ease with which this vital tissue of the tooth may be breached during dental operative procedures. The rich blood supply of the pulp and its rather wide root canals enable more adventurous treatment of the deciduous pulp than is possible with permanent teeth.

### Further clinical applications of a knowledge of tooth morphology
When teeth are restored after destruction by trauma or disease the pattern of restoration should take into account the protection of the gingiva by the cervical region of the crown. Correct articulation with opposing teeth is important since fillings which are 'high' can cause considerable discomfort. The force of mastication may be borne by the tooth with the high restoration. A thorough knowledge of tooth morphology is thus necessary for the creation of optimal function capability for the restored tooth. A wear facet is found between the teeth where they rub together during chewing. This is known as the contact area and teeth are held in contact mainly by the interdental fibres of the periodontal membrane. If the contact is lost between teeth food tends to pack down between the teeth and damage to the gingivae in that area will result. Proper restoration of the mesial and distal marginal ridges is thus important in protecting the interdental gingiva from this type of food packing.

A knowledge of root morphology is important in the design and choice of forceps used in the extraction of teeth. The forceps should grasp the tooth root and not slide down it. The form of the root determines the movements required to extract the tooth. For example upper incisors require a rotary movement to extract the conical roots but such a movement with the lower incisors is almost certain to fracture the slender root. Many oral surgeons prefer to have a radiograph of the tooth prior to extraction so that difficulties may be foreseen. Root canal treatment is difficult without a knowledge of the form, size and number of the roots of the tooth.

# 4

# Dental tissues

As we have seen a tooth has two parts, the crown and the root. The crown can be defined either clinically or anatomically. The clinical crown is the part which is exposed in the mouth. This may vary from time to time, e.g., just after eruption the part exposed in the mouth may be quite small whereas in old age as the gums recede the exposed part may be quite long. The anatomical crown is defined as the part of the tooth covered by enamel and hence its extent is not dependant on age (Fig. 4.1). The root of a tooth is covered by cementum by which it is attached to the bony socket. The bulk of the tooth underneath the enamel and cementum is dentine with an inner core of pulp tissue.

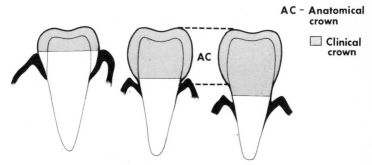

AC – **Anatomical crown**

☐ **Clinical crown**

**Fig. 4.1** Diagram showing the relationship of the anatomical crown to the clinical crown. Note that the anatomical crown is constant

## Enamel

*Macroscopic appearance:* Enamel is the hardest tissue of the body due to its high content of calcium and phosphate. The variation in its colour becomes very obvious when selecting the correct shade of material to restore an anterior tooth. The colour is produced by the degree of translucency of the enamel which is probably related to its degree of calcification. The tip of the tooth often has a bluish white colour. This is because some light passes through without reflection to the oral cavity. Further apically most of the light is reflected from the dentine which lies immediately underneath the enamel giving the enamel a yellowish hue. Teeth which are poorly mineralised have a whiter appearance probably due to irregular crystalline structure of the enamel so that light does not reach the dentine. Whiter teeth are not necessarily better teeth. The colour of a tooth may be affected by disease processes which induce hypomineralisation such as a carious lesion. In the early stage this appears as a whitish patch. Systemic conditions may give rise to a derangement of the development of the enamel producing marks on the surface. These marks are a permanent reminder of the age at which the disease occurred in the individual. The ingestion of excessive fluoride during tooth development results in mottled enamel, which varies in severity from small chalky spots to gross discoloration. This enamel is more resistant to decay at any level of mottling severity, than enamel which is free of fluoride.

The colour of a tooth is affected by material in the saliva, food deposits and habits such as smoking. These 'extrinsic' stains are due to the mucin in saliva adhering to the enamel and providing a means of attachment for many substances. Even with regular cleaning the teeth become yellower with advancing years. This is of great importance aesthetically and elderly patients will mistakenly request whiter teeth when dentures are being constructed, than is compatible with their age.

In children the deposits on the teeth sometimes are of a brown or green colour due to the colonisation with chromogenic bacteria. These deposits can be removed with a polishing brush. In the adult the brown stain of nicotine occurs most frequently on the palatal and lingual surfaces of the teeth particularly in the lower incisor region. Strong tea may also cause brown stains on the inner surfaces of the teeth. One of the anti-plaque agents, chlorhexidine, also causes staining. Hard deposits on the surface of the enamel are known as salivary calculus and along with dental plaque will be further considered in Chapter 6.

*Microstructure:* In order to examine enamel with the microscope ground sections are required. Enamel has such a high mineral content, that decalcification methods used in routine histological preparation remove the enamel completely. Ground sections may be prepared as the name suggests by grinding a tooth first on one side and then the other so that a thin slice of the tooth remains. Another less laborious method is to use a rotating carborundum or diamond disc to slice through a tooth. The slices so obtained are then rubbed down with abrasives until the desired thickness is reached. Sections of between 50-100 $\mu$m are commonly used, as compared with 5-10 $\mu$m for the more usual soft tissue or decalcified hard tissue sections. It is not usual to stain ground sections since hard tissues do not stain readily using routine histological methods.

Using the light microscope the enamel can be seen to be made up of prisms or rods which run from the junction with the dentine to the outside of the tooth. The rods vary in the course that they take depending on the part of the crown in which they lie (Fig. 4.2). Over the tip of the crown the pattern is extremely complex with the rods

**Fig. 4.2** Diagram of enamel to indicate the directions of the enamel prisms. At the tip of the cusp much interweaving takes place. At the sides prisms seem to be criss-crossing in alternate horizontal layers

apparently weaving between each other in a confused pattern. Along the sides of the tooth the rods appear to be in horizontal layers with the rods in adjacent layers running at an angle to each other. Further towards the cervical margin the rods run almost straight out from the dentine to the surface with a slight inclination downwards towards the root.

On cross section the rods again have a variable appearance which does not relate to any particular site on the crown (Fig. 4.3). The commonest appearance is one of crowded hexagons like a bundle of

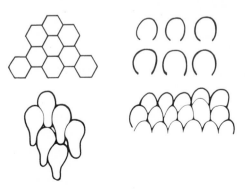

**Fig. 4.3** Various patterns of outlines of enamel prisms in different areas of the enamel

pencils held tight together. Another appearance is of horse shoe shaped rods with the open sides all facing in one direction. They may produce a fish scale appearance or the rod may appear to have a circular head with an extension on one side known as the tail. Along the length of the rod there are numerous varicosities or swellings which produce in ground sections the appearance of cross striations. The planes produced by the array of the rods is the preferred plane for breaks or cracks in the enamel. The edges of a carious cavity often have undermined enamel with unsupported rods. Removal of tooth substance with dental instruments may also undermine the enamel and in these cases the unsupported rods are planed back to prevent subsequent collapse and leakage in the restoration. The rods are approximately 4 μm in diameter so there are literally millions to each tooth. Each rod has an apparent sheath which is thought to contain a higher organic content than the rod itself. Between the rods is a variable amount of interrod enamel which is essentially similar in composition to the material in the rods.

With the electron microscope it has been demonstrated that the appearances seen with the light microscope are due to the arrangement of minute inorganic crystals or crystallites. The crystallites run in the long axis of the rod in the centre but at the periphery they incline further away from the long axis (Fig. 4.4).

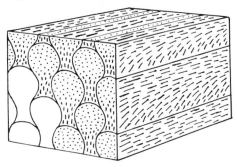

**Fig. 4.4** Diagram to show the pattern of crystallites in the prisms. The boundaries of the prism are demarcated by a change in direction of crystallites (after Meckel, Griebstein & Neal)

Thus at the boundary between prisms the orientation of the crystallites abruptly changes. It is believed that at these place of abrupt change the organic matrix material may be slightly more concentrated. The abrupt changes in crystallite direction in addition to the higher concentration in organic material probably produce the light microscope appearance of rods and their sheaths.

Enamel in ground sections has brown lines running through it indicating the incremental pattern in which it has been laid down. These lines on longitudinal sections run from the amelo-dentinal junction over the cusp or incisal edge to the amelo-dentinal junction of the other side of the tooth. Further cervically they run from the amelo-dentinal junction upwards and outwards to the surface of the enamel. They are known as the brown striae of Retzius, though the reason for the brown coloration is obscure. Where successive layers of enamel outcrop on the surface transverse lines called perikymata are very obvious when the tooth has first erupted. They wear off with age where there is considerable abrasion on the tooth but near the protected cervical margin they tend to persist (Fig. 4.5). On transverse sections of the crown of a tooth the incremental lines are concentric with the amelo-dentinal junction. Spindles appear to originate in the dentine and are continuous with the dentinal tubules.

**Fig. 4.5** The brown striae of Retzius meet the enamel surface at the sides of the crown and cause surface ridges called perikymata

They are most numerous near the amelo-dentinal junction at the tip of the cusp and incisal edge and extend into the enamel for some 10–20 $\mu$m. Tufts and lamellae are sheets of organic material disposed in a vertical plane in the long axis of the tooth. The tufts are found in the inner third of the enamel. As they are vertically placed they are readily seen in transverse sections, where the tufts because of their irregular form look like tufts of grass. The lamellae run through the whole thickness of the enamel and resemble cracks but careful decalcification of a section reveals their true organic nature. They may represent sites of weakness in a tooth and possible ingress sites for bacteria (Fig. 4.6). In older persons the teeth may actually split along a pre-existing lamella.

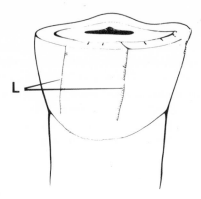

**Fig. 4.6** Enamel lamellae run from the amelo dentinal junction to the surface, in a vertical plane

*Permeability:* Enamel in the young state is slightly permeable and readily adsorbs material from saliva. Fluoride in particular can be adsorbed onto the surface enamel and replace hydroxyl ions in the calcium phosphate crystal lattice. This renders it more acid resistant since fluorapatite is less acid soluble than hydroxyapatite. Enamel also contains citrate and magnesium. In geographical areas with high amounts of fluoride in the water the enamel contains appreciable amounts of this ion, incorporated into the apatite lattice. If the amount of fluoride in the water supply rises above about 4-5 ppm the teeth may exhibit white flecks or spots and in severe cases of fluorosis the enamel structure may be defective. This is unlikely to occur at the concentration of 1 ppm in the water, which is considered to be the optimum concentration that reduces caries incidence without risk of affecting the colour of the tooth. The total inorganic material amounts to about 96 per cent by weight. Organic material and water make up the remaining 4 per cent. The organic material is difficult to collect for analysis and some doubt still remains as to its exact nature. It does not contain the characteristic amino-acids of collagen but does have some affinities with keratin.

The surface enamel is harder and less soluble than the subsurface enamel. This surface zone may be prismless in places though evidence with the scanning electron microscope indicates that there is great variation in its appearance over individual teeth.

In recent years acids have been used to create a 'key' in the surface enamel onto which plastic resins can be bonded. The acid leaches out part of the enamel prism, selectively dissolving either the prism core or the surrounding 'sheath'. The bonding of the resin is used for prevention of caries by filling up deep fissures where stagnation might otherwise occur, or for providing a good seal and better retentive power for a filling. The resin is also used in orthodontics for attaching appliances to the tooth. Etched enamel not covered by plastic resin often 'repairs' itself, probably by remineralisation from the saliva. Apart from this remineralisation, enamel has no capacity for repair.

## Dentine

This tissue makes up the bulk of the tooth. It meets the enamel at the amelo-dentinal junction which on surface view has the appearance of beaten copper with shallow concavities facing towards the enamel. Its yellow colour transmitted through the translucent enamel helps to give the tooth its colour.

Dentine has millions of tubules running through it from the pulp to the outer surface of the tissue. The dentinal tubules originally all contain protoplasmic extensions of cells, known as odontoblasts which line the junction of the dentine with the pulp. The intertubular material of dentine is very similar in its structure to the intercellular material in bone. It contains collagen, ground substance and an inorganic element mainly hydroxyapatite. The dentinal tubules run parallel to each other and are much less complicated in their course than the rods of the enamel. Under the cusps and incisal edges the dentinal tubules run more or less vertically. At the sides of the crown the tubules have an S shape with the first convexity towards the crown and the inner termination of the tubules being at a more rootwards position than the outer end (Fig. 4.7). Nearer the apex of the root the dentinal tubules are more nearly horizontal. The dentinal tubules have numerous irregular small side branches which often appear to meet up with neighbouring tubules or branches. Towards the periphery the branches become larger and the tubules may end by dividing into two.

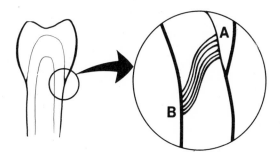

**Fig. 4.7** The dentinal tubules have a primary curvature which is convex upwards at the periphery and concave upwards in its inner half. The odontoblast cell bodies are much further apically than their extremities at the amelo-dentinal junction

In the intertubular zone the collagen matrix is composed of a dense feltwork of fine fibres, orientated in a plane perpendicular to the tubules. At the periphery the collagenous matrix is made up of rather coarse fibres which run parallel to the tubules.

At its junction with the pulp there is a narrow uncalcified zone of dentine, the predentine. Dentine is produced by the odontoblasts throughout the life of the tooth and predentine represents the most recently formed matrix. Calcification occurs as an extension from the dentine already calcified, in the form of small spheres or globules. These globules meet up with each other and normally fuse. Minor defects in this fusion process occur and uncalcified spaces, called inter-globular spaces are found in the crown (Fig. 4.8). They are sometimes aligned in such a way that a series appears to have been produced at a specific time, but usually they are not related to the incremental pattern. They are not found in the dentine of the root.

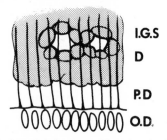

**Fig. 4.8** Diagram of the interglobular spaces. The calcospherites fail to coalesce and an uncalcified irregular space is left. IGS = interglobular space; D = dentine; PD = predentine; OD = odontoblasts

On ground sections just internal to the cementum of the root there is a band of what appear to be small dark granules. These granules are in fact small irregular spaces in the dentine and are called the 'granular layer of Tomes'. They seem to be expanded ends of the dentinal tubules but how they are caused is not known (Fig. 4.9).

Dentine, unlike enamel, can respond to stimuli as it is a living tissue, with its own cells, the odontoblasts. When stimuli of sufficient strength are applied to dentine the odontoblasts lay down extra dentine in the area affected. This extra dentine is called irregular secondary dentine or reparative dentine and is an attempt to wall off the pulp from the source of the stimulation. If the stimulus is too great some or all of the odontoblasts may die and the tubules affected then fill up with gaseous products of the degeneration of the cell. Dentine formation at the inner ends of the affected tubules may still continue with a reduced number of tubules or even none at all. The gas filled tubules appear on ground sections as a dark zone running from the periphery to the pulp. The dark appearance is due to differences between the refractive index of the gas and the dentine within the mounting medium. Any extra dentine which is laid down at the pulp end of the dentine is confined to the areas that are related to the ends of the affected tubules. Hence a lesion on the side wall of the crown results in reparative dentine further rootwards in the pulp (see Fig. 4.7). This is of importance when caries attacks the tooth.

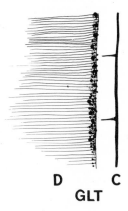

**Fig. 4.9** Diagram of a ground section of the upper part of the root of a tooth. GLT = granular layer of Tomes; C = cementum; D = dentine. The dentinal tubules appear to end in the granular layer

It is also important for the dental surgeon to bear in mind the probable location of reparative dentine. Often he will place a temporary dressing in a deep carious cavity without removing all the decayed dentine. At a subsequent visit after secondary dentine has formed a thicker barrier between the cavity and the pulp, he can then remove the remainder of the decayed dentine and place a permanent filling in the cavity.

Attrition or wear on the occlusal or incisal surface can sometimes cause the loss of enamel and expose the dentine. This results in deposition of reparative dentine in the pulp horns or in the roof of the pulp chamber. If attrition is severe it may eventually expose the reparative dentine.

The tubules of dentine are normally of about 1.5 $\mu$m in width. Within the tubule a lining of peritubular dentine, more highly calcified than the intertubular dentine, accumulates slowly during life. The tubule may be almost or even completely filled by it and the process has been likened to the accumulation of furr in water pipes. The process can be accelerated by stimuli of lesser intensity than that which kills off the odontoblast. The dentine in which the tubules are completely occluded is more translucent than normal dentine and is known as 'translucent' or 'sclerosed' dentine. It is brittle and harder than the rest of the dentine. Sclerosed dentine progressively affects normal teeth with age starting at the root apex. The length of sclerosed root is often used as an index to help in age determination in forensic and anthropological studies. While the translucency can be readily seen in teeth held up to the light it is best demonstrated on longitudinal ground sections examined with the light microscope (Fig. 4.10). Dentine also becomes less sensitive, thicker and more brittle with age.

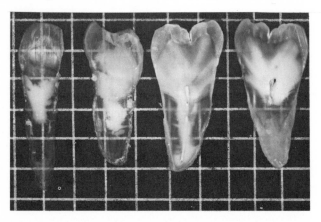

**Fig. 4.10** Age changes in the dentine of the root render it almost transparent on sections. These sections of old teeth show varying degrees of progression of this change up the root starting from the apex

*Chemical nature:* The inorganic material is very similar to the inorganic material of enamel though the crystallites are smaller. Fluorides can be incorporated into the apatite lattice as in enamel but due to the lower amounts of inorganic substance in dentine, its protective effect is less. In addition fluoride can only enter dentine to any large extent during formation of the tissue, and topical application to the crown will not reach enamel covered dentine. The organic material of the matrix is high in collagen content but it also contains acid mucopolysaccharides.

*Sensitivity of dentine:* When caries progresses through the enamel of a tooth and reaches the dentine, it rarely causes pain. However the effect of temperature changes or of chemical stimulation, e.g., by sweet substances when the amelo-dentinal junction has been breached, is well known. When the dental surgeon drills through the enamel, vibration is felt but unless the tooth becomes very hot or very cold little or no pain is produced. Instrumentation on the dentine however is accompanied by pain. Dentine which has become exposed at the neck of the tooth can also be extremely sensitive to touch and temperature changes.

In spite of extensive research on this problem the exact nature of the mechanism of dentine sensitivity is not clear. Thus methods to reduce pain in dentine except by interruption of nerve conduction outside the tooth, are to a large extent empirical. Whatever the stimulus the pulp response is pain. Obtundants are materials which reduce the sensitivity of dentine when applied topically. They are nearly all protein precipitants and may well act by precipitation of tubule contents. In order to protect the pulp it is usual to line the dentinal surface of a cavity or of a fractured tooth, with a bland non-conducting non-irritant material, before the restoration is placed.

## Cementum

This is a bone like substance which covers the roots of the teeth. The cementum provides an attachment for the fibres of the periodontal membrane and by this means the tooth is attached to the alveolar bone of the socket. Cementum normally covers the root portion of the tooth completely. At the neck of the tooth it is quite a thin layer but at the apex of a tooth or between the roots of multi-rooted teeth it is a much thicker layer. It becomes exposed in the mouth when the ginigivae recede due to age or disease, or sometimes as a result of periodontal surgery. This exposed cementum has poor resistance to abrasion. When it is lost the underlying dentine becomes exposed and the tooth becomes sensitive. Note that it is the exposed dentine which is sensitive and not the cementum.

*Structure:* Essentially this tissue has a similar structure to bone over much of the root. In the cervical third of the root however cells are not incorporated in the cementum and thus it is known as acellular cementum. As this is the first cementum to be laid down it is also known as primary cementum. The remaining cementum is known as secondary or cellular cementum.

*Primary cementum:* This appears in ground sections as a translucent almost structureless layer about 100 $\mu$m thick lying outside the dentine in the cervical third of the root. This layer may contain lines running parallel to the root surface representing incremental periods in the production of cementum (incremental lines of Salter). Lines perpendicular to the surface seen in section are the sites of insertion of fibrous bundles from the periodontal membrane. Internal to the cementum is the granular layer of Tomes but the boundary between cementum and dentine is not always clearly delineated.

The junction of cementum and enamel is often described in text books as being of three types. In the first type the enamel meets cementum at a butt joint. Secondly the enamel is overlapped by cementum and in the third type there is a gap between enamel and cementum. Recently it has been shown with the scanning electron microscope that on any one tooth the junction is very irregular and all three situations exist around the neck of a tooth (Fig. 4.11). This may explain the sensitivity that is felt when the root of the tooth is exposed in the mouth, as dentine in the 'gap junction' is exposed even without any abrasion of cementum. Clinically it is often noted that teeth are not equally sensitive all around the neck.

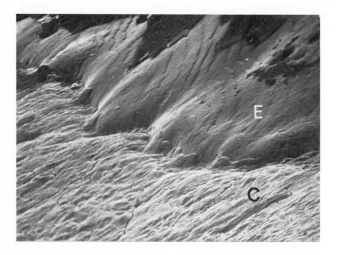

**Fig. 4.11** The cemento-enamel junction seen with the scanning electron microscope. Note how irregular this junction is on this tooth. E = enamel; C = cementum

*Secondary cementum:* This type of cementum is found in the apical third of the root and may overlie a layer of acellular cementum in the middle third. It also is found at the bifurcation region of the roots in molars. Secondary cementum contains cells called cementocytes arranged with their long axis parallel to the surface. They lie in lacunae with canaliculi extending from the lacunae towards the periodontal membrane. The arrangement of the cells is more irregular than in bone but occasionally they appear to form layers. No Haversian systems are found and the cementum is entirely avascular. Resting or incremental lines, which in general are parallel to the surface of the root are similar to those in bone. The cementum is thickest at the apex of the root, and increases in this region with age. When teeth undergo attrition or when tooth movement occurs extra secondary cementum is deposited on the root.

The matrix of both primary and secondary cementum contains collagen which is arranged as intrinsic and extrinsic fibres. The intrinsic fibres are calcified to the same extent as the collagenous matrix in bone. The extrinsic fibres are the ends of bundles of collagen fibres entering the cementum from the periodontal membrane. They are calcified to a variable extent around the periphery of the bundle and are called Sharpey's fibres.

The inorganic material in cementum is mainly calcium phosphate in the form of hydroxyapatite crystals. These crystals have a similar size and arrangement in relation to the intrinsic fibres as is the case with bone matrix.

*Function of cementum:* The cementum provides a means of attachment of the fibres of the periodontal ligament. When a tooth moves through the alveolar bone, either physiologically or by orthodontic forces, cementum deposition enables the periodontal membrane to achieve new attachment by the incorporation of more extrinsic fibres.

Cementum deposition around the apex of a tooth occurs more rapidly where there is a considerable degree of attrition on the biting surface of the tooth. Resorption of cementum may occur in localised areas of the root in permanent teeth and is the usual accompaniment of shedding deciduous teeth.

## Pulp

A tooth has a living soft tissue core, the pulp. This is divided into a pulp chamber in the crown and a central canal in the root. In multirooted teeth there is a canal in each root and even in single roots sometimes the root canal may be duplicated. The pulp chamber tapers towards the root canal but there is no abrupt transition between the two. In some teeth extensions of the pulp are found under the cusps. These are known as cornua or pulp horns (Fig. 4.12). They must be

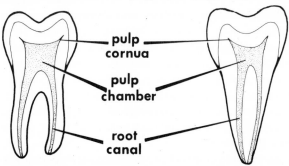

**Fig. 4.12** Diagram of the pulp of a tooth in section. The cornua of the pulp are very vulnerable in cavity preparation especially in deciduous and young permanent teeth

borne in mind when operative procedures are planned on the teeth. In deciduous teeth the pulp chambers are relatively larger than in permanent teeth.

*Structure:* At the periphery of the pulp the odontoblasts form a layer below the predentine. They are tall cells arranged in an irregular fashion, each cell sending a process into the dentine. Amongst the odontoblasts there is a dense capillary plexus, which is supplied and drained by parent blood vessels in the centre of the pulp. In this area also there is a rich plexus of nonmedullated nerves. Some branches of this plexus end on odontoblasts and some may even enter the dentinal tubules.

The pulp is very soft in consistency and has been likened to the loose jelly like material of the umbilical cord. In the young pulp there is a rich cellular content, the cells comprising in addition to the odontoblasts, fibroblasts, histiocytes and wandering macrophage cells. There are also cells which appear to be undifferentiated cells. They are thought to be capable of maturing into fibroblasts or even odontoblasts in the appropriate environment. Lymphatic vessels are also found in the pulp, and they along with the small arteries and veins pass through the small apical foramina to reach the periodontal membrane. Occasionally, there are accessory root canals which open on the side of the root or at the bifurcation of multirooted teeth. Sometimes in operations to remove the pulp tissue they give rise to complications. Since the pulp chamber is enclosed almost completely by hard tissue any inflammation in it causes an increase in pressure. This pressure in addition to producing pain can result in strangulation of the apical vessels leading to death of the pulp. A common form of treatment for this condition is extirpation or removal of the pulp tissue. A knowledge of the anatomy and extent of the pulp cavity and root canals in each tooth is an obvious prerequisite in root canal therapy.

*Reactions of the pulp:* Chemical and thermal stimuli give rise to reactions in the pulp. There may be secondary or reparative dentine production or the pulp may become inflamed. Although pulp death or necrosis often occur it does not invariably follow and at times the pulp may recover. Lining cavities with a bland non-conducting material has already been mentioned as a means of protection of the pulp from thermal or chemical stimuli. Exposure of the pulp to the saliva results in infection of the pulp and care must be taken to avoid this occurrence if possible.

With age the pulp chamber becomes smaller by deposition of more dentine on the roof and walls. Although dentine production goes on through the life of the tooth it is a slow process. The jelly-like ground substance of the pulp is progressively infiltrated by collagenous fibres and the pulp becomes less cellular. The number of nerves is reduced and the tooth becomes less sensitive. Irregular calcifications in the pulp (pulp stones) may occur at any age but the incidence of these changes increases with age.

Occasionally the pulp may die without any symptoms and it is difficult to find the cause. This is more common in upper anterior teeth where forgotten trauma may have played a part. Dead teeth often darken in colour and this may be the only complaint of the patient. The dark colouration is probably due to breakdown products of blood pigments gradually permeating the dentine.

# 5

# Tooth attachment

It is commonly believed that more teeth are lost through periodontal disease than through dental decay. With the present trend of preventive dental care it is likely that people will retain their teeth longer than at present. This will result in a greater emphasis on the treatment of periodontal disease. It is therefore necessary for the dental student to have a thorough basic knowledge of the attachment apparatus so that measures to combat this disease may have a rational basis.

The teeth are implanted in the jaw bone in sockets known as alveoli. Between the bone and the tooth there is a gap approximately 0.2–0.3 mm wide, which is filled with the periodontal membrane. This membrane, sometimes called a ligament, provides a soft cushion for the forces of mastication acting upon the teeth. As we have seen, the fibres of the periodontal membrane are attached to the root by means of the cementum.

### The alveolar bone
The bone lining the tooth socket is also known as the lamina dura (hard sheet) because it appears as a dense white line on radiographs of the teeth *in situ* (Fig. 5.1). The lamina dura is in reality a sieve like structure and on the dried skull it can be seen to have numerous perforations through which the periodontal space communicates with the cancellous spaces in the bone. Histologically the lamina dura is made up of lamellar bone which contains in places Haversian systems, thus meriting the name 'compact bone'. This is not always the case, and the lamina dura is sometimes quite thin. Around the mouth of the socket the alveolar bone may contain woven bone especially if the tooth is being moved by orthodontic or natural forces.

Collagen fibres from the periodontal membrane are inserted into the alveolar bone in bundles, which give it the name 'bundle bone'. These embedded fibres are known as Sharpey's fibres. The bone here has therefore two fibre systems within its matrix, intrinsic fibres which are part of the inherent intercellular calcified substance and

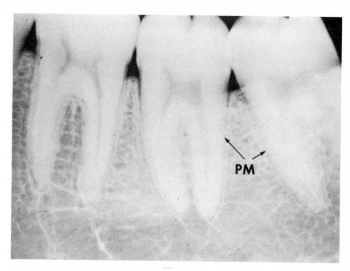

**Fig. 5.1** Radiograph of lower molars. The periodontal membrane (PM) appears as a dark line around the root with the dense lamina dura outside it

extrinsic fibres composed of the bundles of collagen fibres from the periodontal membrane. These latter fibres do not become calcified to the same extent as the intrinsic fibres.

As we have seen cementum also has intrinsic and extrinsic fibres and in structure and composition resembles the alveolar bone very closely. A fuller description will be found in Chapter 4.

### The periodontal membrane

*Fibres:* From the root bundles of collagen fibres run to the lamina dura, the gingival tissues and the adjacent teeth. The bundles, surrounded by a looser tissue in which run the blood vessels, nerves and lymphatics, form the 'principal' fibres of the membrane and are classified according to their position and attachments (Fig. 5.2).

The *oblique fibres* make up the majority of the fibres and pass from the bone obliquely to the cementum in an apical direction along the length of the root. In addition to their apical direction the fibres can be seen in a transverse section of the periodontal membrane to run more tangentially than radially around the root, resembling the spokes of a bicycle wheel (Fig. 5.3). These fibres resist occlusal forces transferring the pressure loading into a tension force on the bone. By their slight tangential arrangement they also resist rotational forces on the tooth. Thus these oblique fibres protect the membrane from being crushed against the bone when loading is applied.

The *apical fibres* run from the apex of the root of the tooth radially to

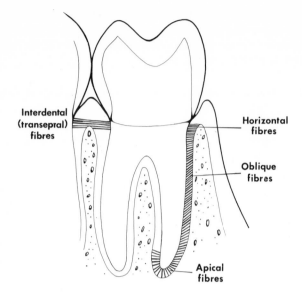

**Fig. 5.2** Diagrammatic view of the principal fibres of the periodontal membrane

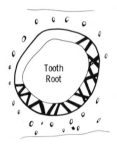

**Fig. 5.3** Transverse section through the periodontal membrane. Only half of the bundles of fibres are shown. Rotational forces are resisted by the arrangement of these fibres

the surrounding bone. They form a cushion through which blood vessels and nerves run to the apical foramen in the root which leads to the root canal. These fibres are broken up when an apical abscess forms around the root.

*Interdental fibres* cross over the interdental bony septum from each tooth to its neighbour. They help to keep the crowns of the teeth in tight contact with each other. When a tooth is extracted new fibres form subsequently linking the teeth adjacent to the gap. As this fibrous tissue contracts during repair, it exerts a force which tends to tilt these teeth towards the gap. The nature of this force is in some doubt but it is believed that fibroblasts are involved in the mechanism. The teeth on either side of the gap may sometimes move bodily towards each other and although it is more common for teeth to move mesially, occasionally premolar teeth may drift distally when the adjacent molar has been extracted.

Fibres of the periodontal membrane which run horizontally from the tooth to the bone at the crest of the socket are called the *horizontal fibres*. They help to prevent any tilting tendency and also aid in keeping the tooth in its socket.

At the bifurcation of the roots in multirooted teeth *interradicular fibres* pass from the bone fan-wise to the surrounding roots. It is a common misconception that these fibres run from one root to another root of the same tooth in this region. A moments reflection will enable the student to see that this arrangement would fulfil no useful function.

The fibres around the neck of the tooth are conveniently described by their attachments but essentially the fibres from the cementum in the cervical region of the tooth and from the crest of the alveolus run to all the surrounding tissues (Fig. 5.4)

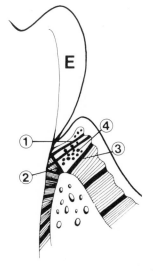

**Fig.** 5.4 Gingival fibres of the periodontal membrane:
1 = dento-gingival;
2 & 3 = alveolar crestal;
4 = circular

*Gingival fibres* run from both bone and tooth to the gingiva. Those from the tooth help in preventing displacement of the gingiva during mastication. They also help in the maintenance of a seal around the neck of the tooth. Bundles of fibres run circumferentially around the tooth in the free gingiva, that is the gingiva which lies adjacent to the tooth and which is not attached to bone, and these are called the marginal or circular ligament.

The gingival and transeptal fibres provide a physical barrier to the invasion of bacteria and are the first fibres to be destroyed by disease processes in pocket formation. In this region of the periodontal membrane there is a continual struggle going on between bacteria and their products on the one hand and the immunological cells, collagen fibres and fibroblasts on the other. Bacterial toxins and antigens can penetrate the epithelium and start the breakdown of the periodontal membrane, hence the importance of preventing the accumulation of bacteria in the cervical region of the tooth. There are no elastic fibres in the periodontal membrane apart from those in the walls of the blood vessels. *Oxytalan fibres* are thin strands which run parallel to the root and end usually on blood vessels. They are relatively scarce and may play a part in the support of blood vessels during loading of the periodontal membrane. They have affinities with elastin.

## Tooth mobility

During chewing the individual teeth move independently of each other in their sockets. This movement is permitted by the undulating nature of the collagen bundles in the periodontal membrane at rest. Pressure on the tooth causes a straightening of these bundles in the initial movement of the tooth into its socket. As the fluid in the tissue spaces is pressurised it escapes through the numerous openings in the socket wall and a second slower intrusion occurs. When the stress on the tooth is removed the fluid is allowed to come back into the periodontal membrane and bring the tooth back to its resting position. Lateral movement of the tooth is similar but the fluid is displaced from the side towards which the force is directed, to the opposite side. Heavy stresses on the teeth are accompanied by elastic deformation of the bone of the socket. Slight deformation of the mandible as a whole often occurs during unilateral chewing.

Gingival inflammation usually has little effect on tooth mobility but changes in the deeper tissues are accompanied by loosening of the teeth. A simple test of the progression of periodontitis is to try to move the tooth with the fingers or with a dental instrument. If there is mobility the degree can be noted to assess on subsequent occasions the progression of disease or the response to therapy. Tooth mobility is also increased during pregnancy. This may be the result of a generalised laxity in collagen fibres in response to circulating hormones. The advantages of such laxity in ligaments in the pelvis during pregnancy and parturition is obvious and the tooth mobility rapidly returns to normal when pregnancy ends.

The independent mobility of the individual teeth results in wear at the contact points. This wear allows forward movement of the teeth through the alveolar bone and in time the approximal contact points become contact areas.

## Blood and nerve supply

The blood supply of the periodontal membrane is from three sources, the vessels to the pulp of the tooth, the vessels from the alveolar bone and the vessels which supply the gingiva (Fig. 5.5). The main direction of the arterioles and venules is longitudinal but there are numerous cross connections forming an extensive network. The multiple connections with the vessels in the bone marrow pierce the lamina dura in numerous places making it sieve-like. When a tooth is extracted there may be considerable loss of blood from the large number of torn blood vessels.

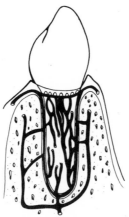

**Fig.** 5.5 Diagrammatic illustration of the blood supply to the periodontal membrane. Many vessels pierce the lamina dura, some anastomose with the gingival vessels and branches are given off by vessels to the pulp

The nerves to the periodontal membrane are branches of the alveolar nerves which supply the pulp of the tooth. They are of two types, nonmyelinated and myelinated. The nonmyelinated are associated with the control of the blood vessels. The myelinated fibres serve a proprioceptive function. They end around the principal fibres in small expanded endings. Pain and pressure are the only sensations appreciated in the periodontal membrane. The pressure endings have a range of response from very light touch to heavy masticatory stress and play a regulatory role in mastication. The sensory nerves are branches of the trigeminal nerve and are inactivated when the pulp of a tooth is anaesthetised during dental treatment. Care should be taken therefore to ensure that occlusal loading is not excessive whilst the teeth are locally anaesthetised. This may occur with restorations which are too 'high' resulting in damage to the periodontal membrane and considerable pain when the local anaesthesia has worn off.

### The cells of the periodontal membrane

In keeping with the densely fibrous nature of the membrane the predominant cell is the fibroblast. These cells are responsible for the laying down of new fibres to meet constantly changing conditions and to provide new attachments when teeth move within the bone. There is evidence that these cells are also responsible for the removal of old fibres.

Lying against the alveolar bone are osteoblasts and osteoclasts which are active in remodelling of the socket. On the root especially in young teeth, cementoblasts line the surface. Defence cells such as histiocytes and macrophages are also found scattered throughout the looser connective tissue.

On longitudinal sections of the periodontal membrane isolated clumps of epithelial cells are found close to the cementum surface. These small groups of cells are the remnants of the epithelial tissue that induced the formation of the root of the tooth and are called 'cell rests of Malassez'. They may proliferate and form cysts in inflammatory conditions of the periodontal membrane. On sections tangential to the root the apparently isolated clumps of cells can be seen to form a slender network close to the cementum.

### The gingiva

The macroscopic appearance of the gingiva has been described in Chapter 2.

The gingiva is divided into the free gingiva and the attached gingiva.

The free gingiva extends for about 1.5 mm from the gingival margin and is marked off by a shallow groove which follows the line of the cervical margin of the tooth. The gingival margin or crest is the rim of a shallow trough or sulcus surrounding the tooth. The trough normally about 1–2 mm in depth contains a fluid which seeps through the epithelial lining of the outer wall of the trough. The base of this trough is the epithelial attachment to the enamel. Between the teeth the free gingiva rises up to fill the space and forms the interdental papilla. The papilla has two more or less discrete peaks, buccally and lingually with a col between them (Fig. 5.6). The epithelium of the free gingiva is either para or nonkeratinised on its outer aspect and nonkeratinised on its inner aspect. The gingival fibres from the tooth and alveolus together with the circular ligament make up the core of the free gingiva. The attached gingiva is covered by keratinised epithelium. It has a stippled appearance due to the bands of collagen fibres which run at right angles to the surface from the lamina propria to the bone. The periosteum merges with the lamina propria and thus the attached gingiva is a mucoperiosteum. The boundary of the attached gingiva is marked by the mucogingival line. At this boundary the mucosa becomes less firmly bound down and contains much more elastic tissue. The epithelium is thinner and nonkeratinised and hence it has a deeper red colour than the gingiva. The width of the attached gingiva varies in different parts of the mouth. It is highly resistant to the abrasive forces of mastication and a reduction in its width often heralds periodontal disease. Periodontal therapy is often of benefit in maintaining or increasing the width of the attached gingiva.

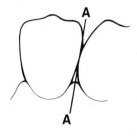

Interdental Papilla

Section through A - A

**Fig. 5.6** The interdental papilla fills the space between the teeth and below the contact area. It is often shaped like a col between two peaks. The lower illustration is the section through A-A

Gingivitis is a very common, almost universal condition which usually starts in the interdental papilla. If severe there is a loss of stippling of the attached gingiva due to oedema, a rolling of the gingival margin and an increase in the amount of fluid seeping into the gingival crevice.

## The epithelium of attachment
The continuity of the oral mucosa is interrupted where the tooth

penetrates the gum. This breach is protected by two mechanisms, the attachment of the gingival fibres to the root of the tooth and the epithelium of attachment. The epithelium is continuous with the epithelium lining the gingival crevice and is sometimes called the junctional epithelium. It is derived from the reduced enamel epithelium which covers the crown of the erupting tooth (Fig. 5.7), though proliferation and mixing of its cells with those of the adjacent gingival epithelium make it difficult to differentiate between the two. The cells produced by this proliferation are shed into the gingival crevice. The junctional epithelium at its connective tissue interface has a basal lamina with hemidesmosomes just as any other epithelial connective tissue junction. On the enamel side there is also a basal lamina between the cells and the enamel and the cell membranes bear hemidesmosomes related to this basal lamina.

The epithelium of attachment extends coronally from the cemento-enamel junction for about 2 mm and with age it tends to migrate down the root of the tooth. It is debatable whether this occurs in the absence of pathology but it is a common occurrence. A basal lamina is then found between the cells and the cementum.

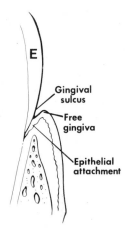

**Fig. 5.7** Diagram of section through the gingival sulcus. In a histological section the enamel (E) would have been lost in the preparation of the section and the gingival sulcus would appear wider than it normally is. The epithelium of attachment is shown here as extending onto the cementum and with age it may be found entirely on this tissue

## Clinical considerations

By the presence of a thin zone of a fibrous, vascular, innervated tissue between tooth and bone, the forces of mastication can be cushioned and controlled. The downward force is transferred to the alveolar bone by the oblique fibres and the hydraulic effect of the tissue fluid protects the fibres from damage. The nerve endings in the periodontal membrane respond and set up inhibitory reflexes to the muscles of mastication. These reflexes together with the information from sensations arising in the oral mucosa, the muscles and temporomandibular joint play a role in the chewing cycle. Reconstruction of the socket during growth by the action of the formative cells of the periodontal membrane allows teeth to migrate, for example in mesial and occlusal drift. The orthodontist makes use of this ability when he exerts force upon the teeth to move them in the correction of malocclusion.

The attachment of the periodontal membrane fibres is not the major factor to be overcome in extracting a tooth. They are readily broken down if the necessary force is applied correctly. It is the close adaptation of the socket to the shape of the root which presents the main retention. Although dilation of the socket is often necessary during extraction of teeth, excessive and uncontrolled force can lead to fracture and loss of part of the alveolar bone.

Infections in the pulp of a tooth can only spread outwards into the apical region and an apical abscess is a common sequel to pulpitis. This may show up as a break in the lamina dura on radiographs as the inflammatory process destroys the bone of the socket. In the upper jaw such infections can spread to the facial surface of the maxilla. From the lateral incisor and the palatal root of the first molar infections tend to spread toward the palate. Occasionally the antrum may be involved from the second premolar and first molar as the roots of these teeth are closely related to it. In the lower jaw an infection from the pulp or periodontal membrane may penetrate the buccal or lingual alveolar plate, depending on the relationship of the root to the bone. If an abscess points lingually, pus will collect above the mylohyoid if the anterior teeth are involved and below the mylohyoid if the apex of the tooth lies below the mylohyoid line. These infections are dangerous as potential spread to the fascial planes around the larynx is threatened.

On the buccal side infections from either upper or lower jaw may reach the superficial loose connective tissue of the cheek and spread widely. The more usual occurrence is for the pus to form a localised swelling of the gum—the so-called gum boil.

The seal between oral epithelium and the tooth is easily damaged and infections sometimes reach the periodontal membrane by this route. The alveolar bone may then be resorbed and as the area of tooth attachment is reduced, so the forces of mastication may become too great for the remaining periodontal membrane. Mobility of the tooth results. Chewing on the affected tooth causes bacteria to be pumped into the periodontal space leading to further inflammation and bone loss.

Preventive measures aimed at reducing the accumulations of bacteria around the necks of the teeth will reduce the incidence of gingivitis and periodontitis.

# 6

# Saliva and salivation

The lining of the mouth is kept moist by the saliva, a mucus containing fluid which helps to keep it in a healthy state. The saliva is not produced by the lining but comes from glands in the underlying or adjacent tissues. In this chapter we shall examine the source of the saliva, its composition and function, and the factors which control its flow rate.

## Sources of saliva

Whole saliva is a complex fluid which comes from the salivary glands, major and minor, and from fluid which accumulates in the gingival crevice or pocket. The major salivary glands are the parotid, the submandibular and the sublingual glands. The minor salivary glands are numerous and are scattered about the mouth. The fluid in the gingival crevice is derived from the tissue fluid of the gingiva.

## The parotid gland

The parotid gets its name from its situation near the 'otic' region where the vestibular and cochlear apparatus is housed. The gland lies between the mastoid process and the back of the mandible and spreads from here forwards along the superficial and deep surfaces of the ascending ramus. The appearance of a formalin fixed gland such as is seen in the dissecting room, gives an erroneous impression of solidity to the gland. In life it is a soft pliable structure readily adapting its shape with the movements of the mandible. The inner part of the gland lies close to the external carotid artery, the external jugular vein and, in front, the medial pterygoid muscle lies between it and the inferior alveolar vessels and nerve (Fig. 6.1). The outer part of the gland overlaps the neck of the mandible and at a lower level it lies on the surface of the masseter muscle. The facial or seventh cranial nerve runs through the superficial part of the gland breaking up into its main branches as it does so. In this part of its course it is a motor nerve and supplies the muscles of facial expression. Operative procedures on the gland must be carried out with care to avoid damaging this important nerve. The excretory duct of the parotid gland which was described by Stenson and bears his name, runs forwards across the masseter muscle before turning inward at the anterior border of the muscle to pierce the buccinator muscle in the cheek. Its opening in the mouth is opposite the upper second premolar. In the living subject the duct may be felt lying on the masseter just below the zygomatic arch.

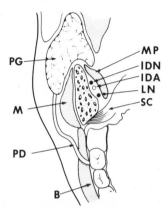

**Fig. 6.1** Horizontal section through the ascending ramus of the mandible and parotid gland. PG = parotid gland; M = masseter; PD = parotid duct; B = buccinator; MP = medial pterygoid; IDN, IDA = inferior dental nerve and inferior dental artery; LN = lingual nerve; SC = superior constrictor

Histologically the gland is made up of spherical or elongated tubular acini which are entirely serous. Thus parotid secretion has a thin watery consistency. The acinar cells have rounded or oval shaped nuclei situated close to the outside or basal part of the cell. The profiles seen on histological section vary from circular to elongated ovoid. In the centre of the acini is the beginning of the duct system and this consists of intercalated ducts, striated ducts and interlobular or

**Fig. 6.2** Diagram of duct system in the salivary gland. ID = intercalated duct; SD = striated duct; ED = excretory duct; OC = oral cavity. The excretory duct is at first interlobular then extraglandular

excretory ducts (Fig. 6.2). Intercalated ducts get this name because they are intercalated or placed between the acini and striated ducts. Striated ducts are so called because the arrangement of mitochondria aligned in rows, together with the folded cell membrane give the base of the duct cells a striated appearance in sections, under the light microscope. These striated ducts are an important part of the gland since the saliva is modified by absorption of some ions here and the secretion of others. The interlobular ducts are lined by columnar or cubical epithelium which has occasional goblet cells. At the oral end of the duct the epithelium becomes stratified and then squamous in character. The opening is guarded by a weak sphincter muscle. This can be dilated using a graded series of silver probes in investigations of the duct system in patients with salivary disorders.

## The submandibular gland

The submandibular gland lies partly on top and partly below the mylohyoid muscle. The two parts are continuous around the posterior border of the muscle. The upper part which is nearer the oral mucous membrane, lies between the mylohyoid laterally and the hyoglossus muscle medially. The deep or lower part of the gland lies between the mylohyoid muscle and the inner aspect of the mandible. A shallow depression on the medial side of the horizontal ramus of the mandible accommodates it and lymph nodes are often embedded in its surface. If one of these lymph nodes becomes enlarged and painful in infections of the teeth or gums it can be palpated with the fingers as a small solid movable mass. The facial artery crosses superficially over this lower part of the gland just before turning over the lower border of the mandible to reach the face. This is a convenient place to monitor the pulse beat particularly during a general anaesthetic in the dental surgery. The duct of the gland runs forward from the anterior border of the upper part raising a fold of mucous membrane in the floor of the mouth and opens on a little papilla to the side of the lingual frenum just behind the lower incisor teeth. It is not uncommon for stones to be found in this duct and it is possibe sometimes to milk small stones from it. The lingual nerve crosses the floor of the mouth under the duct to reach the tongue. Hanging from the lingual nerve at about the middle of its traverse of the floor is the submandibular ganglion, a relay station for the parasympathetic supply to the gland.

Histologically the gland contains acini, 75 per cent of which are serous and the remainder mucus secreting. These are arranged either in separate acini where all the cells secrete mucous or serous saliva or as a central acinus of mucus secreting cells capped by a segment of serous secreting units. These segments appear as small caps partly covering the mucous acini and are called serous demilunes (Fig. 6.3). The secretion from the serous demilunes percolates between the mucous cells to reach the central lumen of the acinus. Mucous secreting cells have flattened nuclei placed at the basal or outer part of the cells. With haematoxylin and eosin staining the cytoplasm of these cells appears clear in contrast to the darker staining serous cells which have a granular cytoplasm. The granularity is due to the refractive index of small granules which contain precursors of the enzymes of saliva and are known as zymogen granules. The histological structure of the duct system in the submandibular gland is similar to that of the parotid.

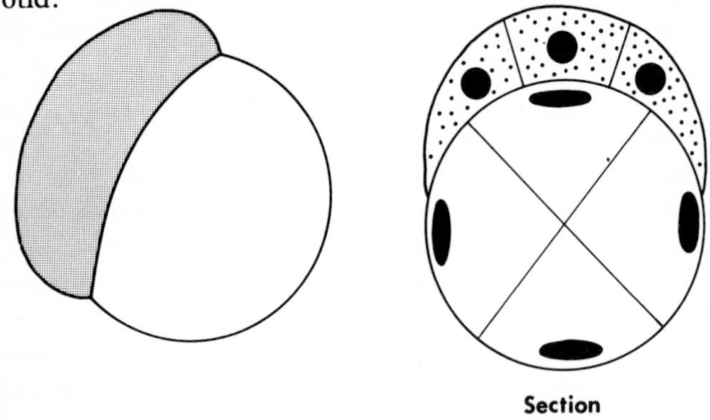

**Section**

**Fig. 6.3** Diagram of a serous demilune. On the right is shown a section through the demilune and its mucous acinus. Note the differences between serous cells and mucous cells. The secretions of the serous cells percolate along the sides of the cells in the central mucous acinus

### The sublingual gland
The sublingual gland is the smallest of the major glands and it lies in the floor of the mouth. It really is made up of a large number of small mucous glands in a loose common capsule. The glands may open into the submandibular duct or separately into the mouth alongside this duct. The mandible has a shallow depression on its medial surface above the anterior part of the mylohyoid line to house the gland. On the inner aspect of the gland are the muscles attached to the genial tubercle, the genioglossus above the geniohyoid below.

Histologically the sublingual gland is almost entirely made up of mucous acini with occasional serous cells arranged in demilunes and in separate acini. There are no striated ducts and since the ducts are shorter than those of the parotid and submandibular glands they have little effect on the secretion of the gland. They are lined by columnar cells with occasional goblet cells.

The sublingual glands are rarely affected by calculi but one of the ducts may become blocked causing the floor of the mouth to swell with the accumulation of the secretions. This is known as a 'ranula' (little frog).

## Minor salivary glands

The minor salivary glands are found deep to the mucosa of the upper and lower lips, the cheeks, the undersurface of the tongue, the soft palate, the dorsal surface of the tongue and the lateral parts of the hard palate behind the first premolar. These minor glands form almost a complete layer in places and are mainly mucous glands. Diseased states of the major glands may affect the minor glands which are more conveniently biopsied to help in diagnosis of such diseases, than the major glands. For example the lower lip is the usual site to biopsy the minor glands in cases of xerostomia or 'dry mouth'. These small glands have short ducts which open directly onto the surface of the mucosa by a multitude of openings. Sometimes one of these little glands become expanded when the duct becomes blocked. This expansion produces a swelling in the tissue and is known as a mucocele or mucous cyst. Rarely does a stone occur in these ducts but the reason for this is unknown.

Histologically the minor glands are made up of mucous acini except for the glands in the tongue which also include serous secreting units. Round the circumvallate papillae, Von Ebner's glands, a group of serous glands, open into and continually wash the trough. Thus tasty substances are flushed out and prevented from remaining in contact with the taste buds for too long.

## The gingival fluid

Fluid from the gingival crevice is a transudate, seeping between the cells of the junctional epithelium. It is not a secretion and is greatly increased when the epithelium is traumatised or diseased. The rate of accumulation of this fluid can be measured by collecting it on filter paper and this measurement has been used as an index of the degree of gingival inflammation.

## Composition and functions of saliva

Analysis of the components of saliva presents many difficulties. Its composition varies from person to person and at different times in the one individual. The relative amounts of each component vary depending on the relative contribution from each source and this in turn is dependant on the stimulus and circumstances at the time. The method of collection affects the analysis, as whole saliva may contain bacteria, epithelial cells and food debris. The bacteria can affect the composition by their metabolism and as the pH also changes on standing, the time between collection and testing can affect the result of analysis. Quantitive measurements of the constituents of saliva are rarely used clinically in diagnosis because of the wide range in values. It is worth noting that saliva, in common with other biological fluids, contains a large number of substances in trace amounts and these have not been estimated.

Water is the main component of saliva in terms of amount present. It acts as a solvent for some of the foods introduced into the mouth and as a diluent to help protect the mucosa from harmful materials such as alcohol. Substances can only be tasted in solution and the water in saliva enables us to appreciate food materials not already in solution.

The main proteins in saliva are the glycoproteins. They have a complex structure with the proportions of protein and carbohydrate in the molecules varying from person to person. They have as their main functional characteristics lubrication and high viscosity. Saliva contains a mixture of these glycoproteins, which were originally called 'mucins' giving rise to 'mucus' in solution. The bolus of food mixed with the glycoprotein becomes a slippery mass easily propelled into and along the oesophagus. Speech is also facilitated by the free movement of the tongue on lips, teeth and palate when all surfaces are liberally coated with the glycoproteins of saliva. Lack of this lubrication gives rise to a 'dry' feeling in the mouth and probably helps in the thirst reflex. The effect of nervousness or anxiey in reducing flow of saliva is appreciated by the feeling of dryness in the mouth and throat. Difficulty in speaking is one result and the provision of drinking water for public speakers is evidence of this function of saliva. Lubrication of the mucosa aids the retention of dentures and helps to protect the mucosa against friction. The glycoproteins in saliva have the property of adsorption to all surfaces in the oral cavity. The coating on the mucosa protects it against the ingress of materials from the saliva. It has been shown in animals for example that carcinogens penetrate the oral epithelium much more easily when salivation is hindered by drugs or surgical intervention. On the teeth the glycoproteins of saliva are associated with pellicle formation, the forerunner of plaque.

The attachment of the pellicle to the tooth is thought to give a foothold for bacterial accumulation. When the pellicle has been colonised in this way it is referred to as plaque. The bacteria in plaque may produce acids by their metabolism of sugars, and the carious process may be initiated. Nearer the gingival margin the plaque bacteria produce toxins which may penetrate the epithelium and initiate gingivitis. Thus the removal of plaque is a preventive measure in both caries and periodontal disease.

Immunoglobulins in secretions from the submandibular and parotid glands are thought to be synthetised in the glands and not diffusion products from the serum. IgA, one of the immunoglobulins in blood is found in relatively high concentration in saliva. The contribution from the gingival crevices is small and edentulous patients show no reduction in salivary immunoglobulin levels. The concept of immunisation against the organisms causing caries has recently been under investigation and the rationale behind this is related to the presence of immunoglobulin antibodies in saliva.

Blood group substances in saliva are of importance in forensic science and in some genetic studies. The agglutinogens A, B and O are found in the saliva of approximately 80 per cent of the population. They are secreted mostly by mucous cells, and in secretors the activity of these substances in saliva is much higher than that of the red blood cells. The substances adsorb readily to tooth surfaces and as they are similar in molecular structure to bacterial coatings they may compete with bacteria for adsorption sites and thus inhibit the attachment of organisms to enamel.

Saliva also contains a protein known as lysozyme which is an antibacterial enzyme. Its activity is not directed equally against all bacteria and it seems to be inhibited by substances in saliva. Other antibacterial factors have been described in saliva though a full scientific evaluation of these components has not yet been carried out.

The only digestive enzyme of any significance in saliva is amylase which is found mainly in parotid and submandibular secretions. It breaks down starch to simpler sugars that are more soluble than the parent starch molecule. Since it acts optimally at pH 7–8 its action in digestion of the food is limited by the small time available while the bolus is in the mouth and the oesophagus, prior to entry to the less favourable acidic conditions within the stomach. Food particles left around the teeth and on the mucosa after a meal are the more likely targets of the salivary amylase.

The inorganic ions, sodium, potassium chloride, calcium phosphate, carbonate and hydrogen are all found in saliva, in amounts which vary with flow rates (Fig. 6.4). The variation in concentration is related to the time that saliva spends in contact with the striated duct cells as some ions are selectively reabsorbed and others are secreted by this part of the duct system. In addition water is absorbed by the duct cells at low rates of flow.

Saliva is hypotonic with respect to plasma. This hypotonicity does not affect the cells of the surface epithelium of the mouth, probably because of the protective effect of the glycoprotein coating. The pH of saliva is not constant, as it is affected by loss of carbon dioxide, the metabolism of bacteria in the presence of various types of food and by the food itself. However bicarbonate and phosphate ions in saliva

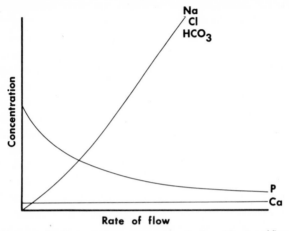

**Fig. 6.4** Patterns of variation in the concentration of saliva with rates of flow (for accurate values see Jenkins's *Physiology and Biochemistry of the Mouth*). As flow rates rise sodium, chlorine and bicarbonate (Na, Cl, HCO3) tend to increase in concentration, while calcium and phosphate (Ca, P) tend to decrease

form fairly powerful buffering systems which tend to keep the pH steady. Thus it is normally possible to ingest fruit juices with low pH values without damage to the teeth. Prolonged contact of these juices may however contribute to dissolution of the enamel. The high incidence of cervical decay in patients with dry mouth is related to the loss of the buffering and cleansing power of saliva. As mentioned above saliva is such a complex and heterogeneous mixture of materials which interact with each other that research on the chemistry and properties of its components has always presented problems. Another complicating feature of the study of salivary composition is the variation with rates of flow. Resting flow rates undergo a 'circadian' rhythm, that is, there is a fluctuation in rate throughout a period of twenty-four hours and this fluctuation is constant from day to day. There is in most people a low resting flow rate early in the morning rising to a peak around noon with a second smaller peak occurring about 5–6 p.m. During sleep the rate drops and salivation may even stop completely. During the daytime commensal oral flora and bacteria introduced into the mouth are constantly flushed towards the pharynx where they are swallowed and finally destroyed in the stomach. This mechanism limits the microbial population in the mouth and hence helps to protect the oral tissues. The lack of the cleansing and antibacterial action of saliva during sleep results in the proliferation of oral bacteria with the consequent unpleasant taste and halitosis which are common after a sleep. Food debris left in the mouth at night provides nutriment for these bacteria. To help prevent dental disease patients should be instructed not to eat after cleaning the teeth last thing at night.

## Control and rates of flow of saliva

Saliva only flows from the salivary glands when the nerves supplying them are stimulated. The glands do not appear to have any inherent activity nor is there any hormonal control as is found in other glands. The parasympathetic and sympathetic systems are involved in the secretion of saliva in a complex manner the mechanism of which is not entirely understood at the present time.

The parasympathetic nerves to the parotid are relayed in the otic ganglion which they reach from the ninth nerve. They are distributed in the auriculo temporal nerve to the gland. The submandibular and sublingual glands receive their parasympathetic innervation from the facial nerve by way of the chorda tympani, the lingual nerve and the submandibular ganglion. The minor glands in the palate are supplied by nerves which relay in the sphenopalatine ganglion. The other minor glands are innervated by parasympathetic nerves arising in one or other of these ganglia but the paths have not been completely established yet. Sympathetic supply to the glands is conveyed mostly in nerves that accompany the blood vessels.

Drugs which have effects on the central nervous system may influence the flow of saliva. For example anti-depressives tend to inhibit the flow rate and dry mouth or xerostomia is one of the side effects of such drugs. Parasympathomimetic drugs such as pilocarpine cause an increase in flow rate from the glands and conversely blockers of cholinergic nerves such as atropine or hyoscine cause a decrease in flow rates. Adrenalin produces an initial increase followed by a decrease in salivary flow.

Salivary flow is greatly influenced by stimuli arising in the mouth. These stimuli are relayed to the pons and medulla thence to the superior and inferior salivary nuclei where synapse occurs with the secretomotor cells. Even without any obvious reflex stimuli however, the glands still secrete though at a very much reduced rate. This probably results from a drying of the oral mucosa which must be kept moist, since saliva even at rest is being swallowed and needs replacing. Taste, especially acidic or bitter substances can elicit as much as a tenfold increase over the resting flow rate. Although individual taste buds adapt rapidly to taste, moving a sapid substance around the mouth stimulates other taste buds and allows the 'adapted' organs to recover. A copious flow occurs with mechanical stimulation of the oral structures and the action of chewing alone causes about a five fold increase over the resting rate, even where the material being chewed is quite tasteless. The presence of objects in the mouth also stimulates salivary flow and dental instrumentation can lead to an embarassingly high rate of flow during dental treatment. Hence the need for salivary ejectors and other aids to keep the operation site dry. Psychic stimuli such as the thought of food may induce salivary flow and emotions such as fear or apprehension tend to reduce secretion of saliva producing a dry and 'thick' sensation in the mouth.

Measurement of salivary flow rates may be useful in assessing the severity of disease of the glands or in providing information in a course of treatment. Whole saliva can be collected by allowing the saliva to drain into a collecting cup or by spitting out at intervals making sure that no swallowing occurs. The latter method does cause some mechanical stimulation and also draws attention to salivation, which may in itself cause stimulation. It is a useful method for measuring the stimulated salivary flow rate. To measure the flow rate of the individual major glands cannulation of the salivary duct or the use of small collecting chambers which fit over the duct openings may be employed.

**Dental disease and saliva**
There have been many attempts to correlate aspects of salivation with dental disease. Since caries is the most prevalent dental disease, especially among children and young adults, this disease has attracted most attention. Caries rate is measured by the number of decayed, missing and filled teeth (DMF index); but the problem is that this reflects *past* disease. Salivary studies reflect *present* conditions. Qualities of saliva that have been examined for a possible role in caries production include rate of flow, calcium and phosphate content, opsonin activity, antibacterial activity in general and against lactobacilli and streptococci in particular, and amylase activity. Only rate of flow and activity against lactobacilli or streptococci show a consistent relationship to the incidence of caries. Caries free individuals, who are rare in westernised societies, tend to have higher flow rates and tend to have smaller number of lactobacilli acidophilus or streptococcus mutans in the saliva. In patients with xerostomia caries of the cervical margins of the teeth is common.

At the present time neither periodontal disease nor the rate at which calculus accumulates has been proven to be related to any aspects of salivation.

# The functional histology of the oral mucosa

In Chapter 2 the structure of the mouth was described from a macroscopic viewpoint. In this chapter the microscopic appearance of the oral mucosa will be considered together with some of the functions of this tissue. The condition of the mucosa is a useful indicator of the state of health of an individual and an understanding of normal function and structure means that disease processes may be dealt with in a rational way.

The oral mucosa consists of covering epithelium and the underlying connective tissue which is known as the lamina propria. For descriptive purposes these are dealt with separately here but it should be remembered that the two tissues are interdependent and the structure of one affects the nature of the other.

### The oral epithelium

The oral epithelium is a stratified squamous layer which is nonkeratinised over most of the oral cavity. The areas that are keratinised are the hard palate, the gingivae and the dorsal surface of the tongue (Fig. 7.1). On the hard palate and the gingivae the mucosa is tightly bound down to the underlying bone, and on the dorsal surface of the tongue to the underlying muscle.

The basal cells are separated from the connective tissue by a basement membrane, and by their proliferation they replace the cells that are lost from the surface. The rate of proliferation of the basal epithelial cells in the mouth is higher than in skin. It has been estimated from animal studies that the oral epithelium can be replaced by division of the basal cells in about ten days. The nonkeratinising epithelium has a slightly higher rate of replacement than the keratinising variety. One important feature of normal epithelium is the property of cell adhesion. The cells remain in contact with each other and when viewed with the electron microscope appear to have small attachment sites called desmosomes between the cells. If the epithelial cells undergo neoplastic or cancerous changes they may lose their attachments and break away from the remainder of the cells.

Entering the blood or lymph vessels they migrate to distant sites and form secondary growths.

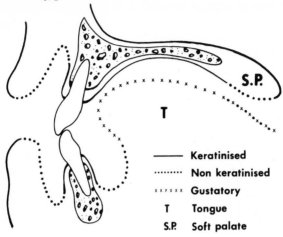

Keratinised
Non keratinised
Gustatory
T       Tongue
S.P.    Soft palate

**Fig. 7.1** Diagram of a sagittal section through the oral cavity to illustrate the variation in the epithelium in different areas. The nasal surface of the hard and soft palate is covered by ciliated columnar epithelium

Normally the control of cell division is thought to come from substances released from the cells in the layers near the surface. These substances passing downwards tend to inhibit cell division as long as the surface layer is intact. If the surface is broken the inhibiting substance (which is known as chalone) does not pass downwards but is released. The basal cells freed from this inhibition proliferate at an increased rate. The surface is then repaired by the differentiation of these rapidly proliferating cells and release of the chalone from the surface stops. As it again accumulates and passes down to the basal cells the brake is again applied to cell production. Another factor thought to be involved in the control of cell division is contact with adjacent cells. When epithelial cells are grown in tissue culture their rate of mitotic activity decreases when a complete layer of cells is formed.

Above the basal cells the stratum spinosum is made up of polyhedral cells with wide intercellular spaces. The spaces are crossed by numerous spinous projections from each cell. With the electron microscope each little spine can be seen to have desmosomes connecting the cells together. The stratum spinosum continues to the surface in nonkeratinising epithelia, the cells becoming flatter and closer together. From the surface the flattened nucleated cells are shed into the saliva. The study of the microscopic appearance of these cells, known as exfoliative cytology is sometimes useful in diagnosis. Buccal smears can also be used to determine the chromosomal sex of an individual.

Keratinising epithelium has two further layers above the stratum spinosum. The first is a single or double row of cells containing dark, keratohyaline granules. This is the stratum granulosum and the granules are thought to be related to the formation of keratin. In the second layer the cells become flatter by the loss of water, the cell membranes become thicker and less permeable and the cell contents more dense. The nucleus is lost and the granules dispersed. This layer is the stratum corneum which stains differently from the remainder of the epithelium (Fig. 7.2).

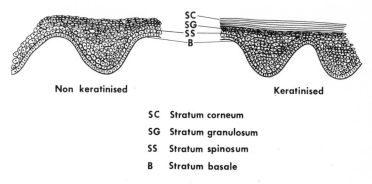

Non keratinised                                          Keratinised

SC    Stratum corneum

SG    Stratum granulosum

SS    Stratum spinosum

B      Stratum basale

**Fig. 7.2** Diagrammatic representation of the two most prevalent types of epithelium in the oral cavity. Basically there are two layers in nonkeratinising and four layers in keratinising epithelium

When the gingivae become inflamed the keratinisation pattern of the epithelium is disturbed. Instead of the normal stratum corneum the cells retain their nuclei, though these nuclei appear small, dark and flattened. This type of change is known as parakeratinisation. Parakeratinisation is also found occasionally in areas on the hard palate without any signs of disease. The keratin layer is often abnormally thick on the palate in subjects who are heavy smokers and sometimes this forms a whitish patch with small red circular spots where the minor salivary glands open onto the surface. The mucosa of the cheek may be keratinised in a line at the level of the occlusal plane. The most probable cause of this keratinised line is trauma during mastication, but the degree of keratinisation is never as great as that found in the palate. Friction applied to gingivae helps to promote normal keratinisation, and the use of wood points between the teeth not only clears away harmful plaque but also stimulates the keratinisation of the epithelium in that region.

## The lamina propria of the oral mucosa

The connective tissue underlying the oral epithelium is known as the lamina or tunica propria. In the region of the lips the lamina propria has prominent papillae or ridges extending into the covering epithelium and these are highly vascular and so the lips appear red. The inner surfaces of the lips and cheeks, the floor of the mouth and the undersurface of the tongue all have a thin elastic lamina propria. In the cheek, bands of fibro-elastic tissue bind the epithelium to the cheek muscle so that it does not get trapped between the teeth when the jaws and teeth are closed. Small mucous glands lie in the deeper part of the connective tissue of the cheek, the lips and the tongue, opening by short ducts onto the surface. In the floor of the mouth the sublingual and submandibular glands also pour their secretions onto the mucosa. Below the lamina propria in some areas, notably the vestibular region a looser submucosa is found. Surgically, in this region the mucosa and the submucosa can be separated from the periosteum.

Over the gingivae and most of the hard palate the epithelium is tightly bound to the periosteum of the bone by dense bands of fibrous tissue forming a mucoperiosteum. This tight binding of fibrous tissue makes it difficult to inject local anaesthetics into these regions without causing considerable discomfort to the patient. At the sides of the hard palate behind the premolars numerous glands separate the mucosa from the bone, and the tissue below the epithelium is much looser. Through this tissue the greater palatine nerve runs to supply the gums on the palatal side of the teeth. Injections in this tissue cause less pain because of the loose connective tissue, and local anaesthesia of the palatal mucosa from the canine to the last molar can be achieved by this procedure.

The depth of the vestibular regions of the mouth is limited by the attachment of the facial muscles to the mandible and maxilla. In the midline and premolar regions the muscles gain a higher attachment and the depth is consequently limited. These high attachments must be avoided in making dentures, otherwise the activity of the muscle will cause dislodgement of the denture and ulceration of the mucosa.

It is worth emphasising that where the mucosa is bound down to bone it is more easily abraded and tends to be keratinised. Where the submucosa is looser and more elastic it yields when force is applied and abrasion is lessened. The surface of the mucosa in such regions is usually nonkeratinised.

## The tongue

On the dorsal surfaces of the tongue there are three types of papillae

(Fig. 2.4). The filiform papillae are conical projections of keratinised epithelium with a narrow core of connective tissue over the major part of the dorsal surface of the tongue. In illness the keratin piles up on the papillae. Normally they provide a rough surface and a loss of this roughness can often be ascribed to systemic disease or to vitamin deficiency. The fungiform papillae are also very numerous especially at the front and the sides of the tongue. On section they resemble mushrooms and are covered by thin, nonkeratinising epithelium so they appear as red spots, among the grey background of the filiform papillae. Along the sides of the fungiform papillae the epithelium is specialised in places forming taste receptors commonly called taste buds. The taste buds are barrel shaped structures with the long axis of the stave like cells perpendicular to the surface of the epithelium. The circumvallate papillae are arranged in a V-shaped line at the junction of the anterior two-thirds and the posterior one-third. There are about twelve of these papillae and they resemble enlarged and flattened fungiform papillae, though they do not rise above the general level of the surface of the tongue. Surrounding each papilla is a trough with taste buds in the epithelium of the outer and inner walls. These taste buds have a structure similar to those associated with the fungiform papillae (Fig. 7.3). Opening into the floor of the trough are small glands which help to wash out the contents after the taste has been appreciated.

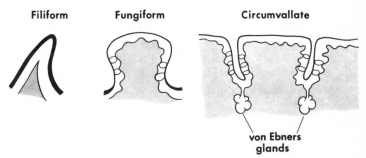

**Filiform**     **Fungiform**     **Circumvallate**

**von Ebners glands**

**Fig. 7.3** The three types of papillae on the dorsum of the tongue. Taste buds are associated with the fungiform and circumvallate papillae

Other structures found sometimes at the side of the posterior one-third of the tongue in the form of vertical folds are the foliate papillae. The recognition of these vertical folds of mucous membrane is important as they can sometimes be mistaken for pathological conditions as stated earlier. Their function is not known.

It is claimed that certain areas of the tongue respond more readily to one type of taste than another. For example sweet taste is appreciated best at the front of the tongue, bitter at the back and sour and salt at the sides. However, there is great individual variation in this pattern. Another theory holds that it is the pattern of impulses from a taste bud nerve which gives a distinct taste. Thus, each taste bud may be receptive to more than one modality.

An interesting aspect of the taste at the back of the tongue is that materials being swallowed often taste differently from their taste at the front of the mouth. Taste buds have been described as being found on the soft palate, the epiglottis and more rarely the pharyngeal wall.

Wearing dentures affects the ability of subjects to detect and identify sapid substances although often patients who have worn dentures for some time appear to become more capable of the identification of tastes presented to them as test substances. The decrease in taste sensitivity in patients with dentures that cover the palate, suggests that taste receptors are present on the hard palate. Dentures when first worn not only reduce taste but lessen texture appreciation. The appreciation of texture is an important aspect of our enjoyment of food and its loss causes much disappointment in those who are first time denture wearers. The oral mucosa also responds to a general chemical stimulation by irritants and hence the effect of curry or mustard or other hot foods. The free nerve endings scattered throughout the oral mucosa are responsible for this sensation. Further consideration of taste will be found in Chapter 12.

### Effects of systemic conditions on the oral mucosa

During pregnancy the gingivae tend to become more liable to inflammation from irritants and pregnancy gingivitis is a condition commonly met in general dental practice. The gingivae become swollen and bleed easily and in some women the teeth show abnormal mobility. The changing balance of hormones released in pregnancy is thought to be involved in the production of this increased vulnerability of the gingivae. Fortunately in the majority of cases the gingivae revert to normal at the end of pregnancy. A similar condition occurs in women taking some types of contraceptive pill. Cyclical changes in the oral mucosa coinciding with menstrual periods have been reported. A type of oral ulceration has a monthly periodicity in women who are susceptible to this condition and these recurring lesions may well have an hormonal basis although direct proof of this is lacking.

Deficiencies of vitamins rarely occur in European countries but a lack of vitamin may be due as much to an inability to absorb it from the gut as a deficiency in the diet. Vitamin A deficiency has been shown to affect the mucosa in experimental animals but there is no evidence for changes in human oral mucosa resulting from deficiency in this vitamin. Lack of vitamin C (ascorbic acid) only occurs in three animals, man, monkey and guinea pig. In man deficiency produces scurvy with bleeding, soreness and sponginess of the gums. The gingival capillaries become enlarged producing swelling and a deep red shiny appearance to the gingivae. Haemorrhage, ulceration and infection may occur. Scurvy is rare nowadays but some cases of gingivitis may be due to a mild degree of deficiency. Vitamin C has been prescribed to aid healing after extractions. From animal experiments it would appear that vitamin C is essential to the proper formation of collagen. Riboflavin, a member of the B group is required for tissue oxidation. Deficiency produces a cracking of the lip at the corners (angular cheilosis) and generalised desquamation in other parts of the face and lips. The tongue becomes magenta coloured and smooth with a reduction in size of filiform papillae. Deficiency of nicotinic acid also affects the tongue making it enlarged, painful and beefy red. The gingivae are painful and the whole mouth may be inflamed. In pernicious anaemia, patients are unable to absorb vitamin $B_{12}$ and a sore smooth tongue is common. With all these vitamin B deficiencies the nature of the changes in the mucosa at the histological level is not understood.

With age the mucosa loses much of its elasticity. It becomes thin and is less capable of withstanding stress. The gingivae tend to recede so that the teeth appear longer, hence 'long in the tooth'. Over the dorsal surface of the tongue deep fissures are found with increasing frequency as age progresses. Small sebaceous glands appear inside the corners of the mouth as yellow spots. They are known as Fordyce's spots but apart from the necessity to recognise what they are, they appear to have no great clinical significance.

### Absorption through the oral mucosa

Absorption through the oral mucosa has long been recognised as a rapid route for medication as drugs absorbed in this way go straight into the general blood circulation. This contrasts with drugs absorbed in the gut which travel in the portal circulation to the liver before reaching the general circulation. By passing through the liver the drug may well be altered and its effect lost, and its concentration in the blood will be slower to reach a peak than with orally absorbed drugs. Examples of drugs where blood levels are required rapidly are trinitroglycerine, which is taken for heart conditions and ergotamine which is used to abort an acute migraine attack. The absorption through the oral mucosa of irritant materials is normally hindered by a coating of saliva over the surface, but this is lost in the condition known as xerostomia or dry mouth. Patients with this condition have abnormally sensitive mucous membranes and complain of burning sensations with acidic and even apparently bland foods like tomatoes. The lack of saliva means that friction is greater when the mucosa rubs against any other object in the mouth. There is increased desquamation of the surface layers and hence increased absorption. As the condition is more prevalent in older age groups where the epithelium may already be thin, it can be seen that many factors contribute to the symptom of 'burning mouth'.

It must be emphasised again that the dental surgeon who sees his patients every six months is in a good position to make an early diagnosis of the causes of slight changes in the oral mucosa and a thorough understanding of the normal condition is essential.

# The development of the mouth

A study of development is very useful as it helps in the understanding of the mature structure. It is also of importance in understanding some disease processes and congenital abnormalities. In addition a knowledge of the timing of various events may lead to a rational approach to the prevention of disturbances of normal development resulting from therapeutic intervention such as drug administration or certain types of dental treatment. Conversely therapeutic measures aimed at the prevention of disease may be administered at the optimum time, e.g., the administration of fluoride supplements during enamel formation to ward off dental decay. General physiological principles are of course also involved. It is useless for example to administer substances to prevent or treat diseases if they do not reach the point of action and a knowledge of the placental barrier function at the various stages of pregnancy is essential for rational treatment.

**General**
The word embryo is loosely used by many people but strictly it signifies the product of conception up to about eight weeks, by which time all the organs have appeared. After this it should be called a fetus and fetal life goes on until birth. The usual guide to the age of the fetus is the crown rump length, since the actual date of conception may not be known. The crown rump length at five weeks is approximately 5 mm. After five weeks the embryo grows about 1 mm every day up to approximately 35 mm. After this growth is slightly more rapid. At the end of three months, i.e., at about twelve weeks the crown rump measurement is approximately 70 mm. At the end of the fourth month this has gone up to 140 mm, the fetus has reached 240 mm by the end of the sixth month and the infant will measure somewhere in the region of 300 mm by the time it is born between thirty-six and forty weeks.

**The early development of the mouth**
When the embryo is about 3 mm long the forebrain is separated from

the developing heart region by a narrow slit. In the depth of this slit is the bucco-pharyngeal membrane behind which lies the foregut. The membrane soon breaks down and continuity between the primitive mouth and the gut is established. Rathkes' pouch is a small diverticulum in the roof of the mouth which grows up to meet part of the brain and help form the pituitary gland.

The upper boundary of the primitive mouth is the fronto-nasal process which becomes divided by a pair of shallow depressions, the nasal placodes, into a median nasal process and two lateral nasal processes (Fig. 8.1). The nasal placodes appear to sink below the surface but this is the result of growth of the surrounding tissue rather than a movement of the placode. Soon the visceral or branchial arches start to appear in the side wall of the embryo between the forebrain and the heart. The first of these arches, the mandibular, grows forward by an extension of its mesoderm and meets the arch of the other side in the midline. The fusion of the two arches forms the lower boundary of the primitive mouth opening, excluding the pericardial region from it.

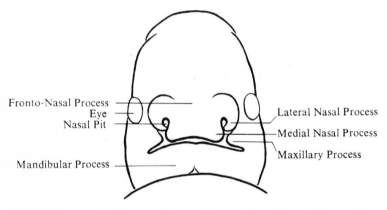

**Fig. 8.1** Diagrammatic illustration of head of human embryo of 10 mm CR length. The mandibular processes have separated the pericardial region from the primitive mouth

From the back of the mandibular arch the maxillary process arises and pushes forward along the upper boundary of the mouth opening. By this time the developing eye has appeared and the maxillary process grows under the depression for the nasal placode converting it into a blind pit. The extension meets the median nasal process, grows in front of it and meets the extension of the maxillary process of the opposite side, thereby cutting off the median nasal process from the upper boundary of the primitive mouth opening (Fig. 8.2). This is now the primary palate as it separates the blind ending olfactory pits from the primitive mouth. The blind pits now deepen and the tissue between them and the back of the mouth breaks down to produce the posterior nares (Fig. 8.3). If the fusion of the maxillary extension is incomplete or breaks down a cleft lip will result. This will lie to one side of the midline and run up towards the nostril. In such a case the median nasal process forms the upper lip in the midline.

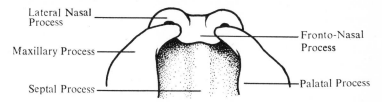

**Fig. 8.2** Diagram of head of a human embryo at about six weeks. The maxillary processes are growing under the median nasal process cutting it off from the upper boundary of the mouth, forming the primary palate

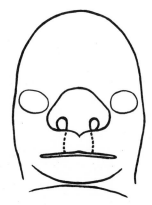

**Fig. 8.3** Diagram of the head of a human embryo of seven weeks. The nasal pits have deepened and open posteriorly into the stomatodaeum. The buried fronto-nasal process is shown in broken line

Inside the primitive mouth three processes project from the walls. In the midline the tecto-septal process hangs down vertically from the roof. This process is continuous in front with the median nasal process and its mesoderm has migrated to this position from the maxillary processes on each side. It will form the nasal septum. The maxillary processes have also given rise to a pair of palatal shelves, one on either side, which hang down at the sides of the tongue (Fig. 8.4). They are in continuity with the primary palate at the front. Between the ninth to eleventh week the tongue descends and the vertically positioned palatal shelves rapidly elevate and meet one another and the nasal septum in the midline (Fig. 8.5). The epithelium at the junctions degenerates and the mesoderm from all three unites. Lack of meeting of the processes, failure of the epithelium to disintegrate or subsequent breakdown after union has occurred all result in various degrees of the deformity known as cleft palate. The causes of clefts are many and X-radiation or surgery during this crucial period of pregnancy should be avoided unless absolutely essential. As some drugs can have a disastrous effect on the fetus medicaments should be prescribed with caution during pregnancy.

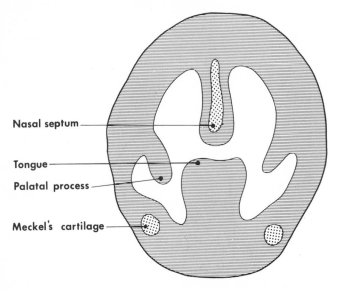

**Fig. 8.4** Diagram of coronal section through the head of a human embryo at about seven weeks. The palatal processes hang down on either side of the tongue

### Development of the tongue
The visceral arches at the side of the head region of the embryo have already been mentioned. The tongue comes from the anterior end of the first arch and from the third arch. From the first arch paired swellings arise and unite with a midline structure the tuberculum impar. From the third visceral arches two swellings unite and grow over the second arch to form the posterior third of the tongue by meeting the anterior part. This junction is seen in the adult tongue as the sulcus terminalis (Fig 8.6). The foramen caecum on the dorsum of the tongue between the anterior and posterior parts, indicates the site of the downgrowth of epithelium which gave rise to the thyroid gland. The track of the epithelium is marked by epithelial remnants which sometimes give rise to cysts in the tongue or the neck.

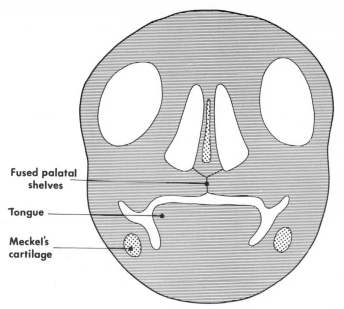

**Fused palatal shelves**

**Tongue**

**Meckel's cartilage**

**Fig. 8.5** Diagram of coronal section through the head of a human fetus at nine weeks. The palatal shelves fuse with each other and the nasal septum. The nasal cavity is thus shut off from the mouth

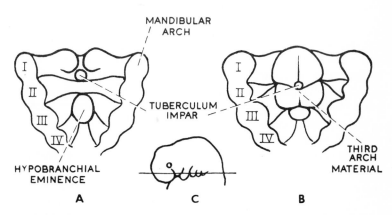

MANDIBULAR ARCH

TUBERCULUM IMPAR

HYPOBRANCHIAL EMINENCE

THIRD ARCH MATERIAL

**A**          **C**          **B**

**Fig. 8.6** Diagram to show the development of the tongue; A. Floor of the mouth at 9 mm CR length. Anterior part of the tongue appearing as paired lateral swellings from the mandibular arch; B. Later stage when third arch contribution has grown over the second arch to form the posterior part of the tongue; C. Plane of section of A and B

The sensory nerve supply of the tongue comes partly from the mandibular nerve (the nerve of the first arch) and partly from the glossopharyngeal nerve (the nerve of the third arch). Special taste fibres are present in the chorda tympani which is from the facial nerve, the nerve of the second arch, and the glossopharyngeal contains taste fibres from the circumvallate papillae. Muscles from the occipital region migrate to the tongue and bring their nerve supply the hypoglossal, with them. Thus the complicated nerve supply of the tongue can be explained by its development.

Movements of the tongue have been detected as early as the tenth week and it is felt that the fetus may perform swallowing movements at this stage just when the palatal shelves are rising from a vertical position to a horizontal position. The tongue may in fact by its muscular activity help to raise the palatal shelves.

### The salivary glands

Between the developing tongue and the mandibular process there is a groove from which the epithelium grows into the underlying connective tissue on either side of the midline. These downgrowths of epithelium consist of solid cords of cells which branch repeatedly to give the acini of the salivary glands. The cords become canalised and form the salivary ducts. In the upper jaw an outgrowth of epithelium commences at the corner of the mouth and extends backwards. It is carried further backwards by the development of the ascending ramus of the mandible and by the development of the masseter muscle. This forms the parotid gland. The gland shows signs of function at about eleven to twelve weeks and the ducts at this stage are tubular.

### Facial skeleton

The early skeleton of the head is the chondrocranium. As its name suggests it is cartilaginous and consists essentially of paired capsules for the middle ear (the otic capsules), a single midline capsule for the nose, the nasal capsule and a basal part connecting these (Fig. 8.7). Much of this chondrocranium is replaced by endochondral ossification but some parts remain until after birth and the nasal septum remains cartilaginous in part throughout adult life. The cartilage of

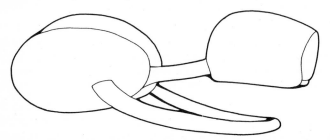

**Fig. 8.7** Diagrammatic illustration of the chondrocranium. The paired otic capsules are joined to the nasal capsule by the basal plate

the chondrocranium is known as primary cartilage. In the visceral arches the first skeletal elements are also cartilaginous. Meckel's cartilage forms in the mandibular arch and runs from the midline to meet the chondrocranium at the otic capsule. The formation of these cartilages starts at about the fifth week.

### The mandible

Within the mandibular process the connective tissue on the outside of Meckel's cartilage starts to form bone in the region of the division of the mandibular nerve into its incisive branch and the mental branch. This bone grows backwards, forwards and upwards on the outside of Meckel's cartilage (Fig. 8.8). As the bone grows backwards along the course of the mandibular nerve on the outside of Meckel's cartilage two small secondary growth cartilages appear. The more important of

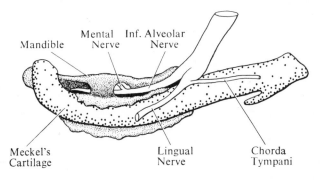

Fig. 8.8 Diagram of the development of the mandible as seen from the lingual aspect. The bridging over of the mental and incisive nerves has commenced

these is the more posterior of the two and this will eventually become the condylar head of the mandible. The other secondary cartilage forms an upwardly growing process called the coronoid process which will eventually give attachment to the temporalis muscle. The condylar process contains a growth cartilage for some twenty years whereas the coronoid process has lost all its cartilaginous element before birth. At the front end of the mandible the two halves unite about one year after birth. Until then they are separated by small pieces of cartilage which are derived partly from Meckel's cartilage and are partly secondary cartilages. The new growths of cartilage in relation to the extension of bone are called secondary cartilage to distinguish them from the primary cartilage of Meckel's cartilage and the chondrocranium. Meckel's cartilage subsequently degenerates and almost completely disappears. The proximal end forms the malleus, one of the tiny ear bones. Between the otic capsule and the mandible it becomes the spheno-mandibular ligament and small portions persist in the symphysis of the mandible until a short time after birth.

### The maxilla

In the upper jaw, bone starts developing in relation to the division of the maxillary nerve into its infraorbital branch and the anterior superior dental branch. From here bone formation proceeds backwards, forwards and inwards towards the midline, downwards towards the developing teeth and also into the fused palatal shelves. Behind the maxilla a centre of ossification starts for the palatine bone which has an upward extension and an inward extension which will eventually form the back part of the hard palate. Secondary cartilages develop also in relation to the maxilla but these are very transitory and play no part in the growth of the maxilla after birth.

The primary cartilage of the nasal capsule gives rise to the ethmoid bone by endochondral ossification of its roof and upper parts of the septum and lateral wall. The inferior turbinate bone also develops in this way. The palatine bone develops on the inner aspect of the posterior part of the nasal capsule and sends a horizontal extension into the back part of the fused palatal shelves. Between the inferior turbinate and the lateral part of the ethmoid the nasal capsule atrophies and an outpouching of the nasal mucosa occurs forming the maxillary sinus. Its extension into the adjoining maxilla hollows out this bone and separates the orbital surface from the alveolar region. At birth the maxillary sinus is very small and is often quoted as being about the size of a split pea. The only part of the nasal capsule that persists into adult life is the anterior part of the septum. The remainder either is replaced by bone or atrophies.

### The temporomandibular joint

The developing mandible comes into relationship with the skull in the temporal region with the development of the secondary cartilage at the condylar head. A strip of tissue intervenes between the two and is attached to the lateral pterygoid muscle. This strip of tissue later becomes the intra-articular disc. Cavities appear above and below this strip of tissue at about the 70 mm stage and at the time of birth the joint cavities are relatively large but the articular eminence hardly exists. The articular slope develops as the deciduous dentition erupts.

### The development of the teeth

*Early development*
The epithelium covering the mandibular and maxillary processes at

about six weeks consists of two layers, a basal layer of cuboidal cells and an upper layer of rather larger and more irregular cells. They are separated from the underlying mesodermal tissue by the basement membrane. Around the crest of the arches a horseshoe shaped line of proliferation occurs sending a vertical projection into the mesoderm. This is the primary epithelial band and it rapidly splits into an outer vertically inclined process and an inner horizontally directed band. The outer band is the vestibular lamina and its epithelium will eventually break down in the central region to form the vestibule of the mouth, separating off the lips and cheeks from the jaws (Fig. 8.9).

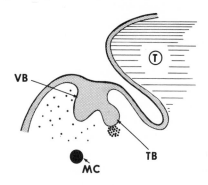

**Fig. 8.9** Diagram of a section through the lower jaw showing the vestibular band and the tooth band as ingrowths of the epithelium. VB = vestibular band; TB = toothband; MC = Meckel's cartilage; T = tongue

The inner band is the dental lamina and on it the tooth germs will develop. Along the dental lamina small rounded swellings are created by localised mitotic activity at a higher rate than in the neighbouring areas (Fig. 8.10). These represent the definitive enamel organs of the deciduous teeth. As they appear the mesenchyme immediately below

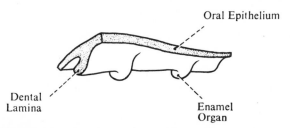

**Fig. 8.10** Diagram of a strip of oral epithelium with part of the lower dental lamina. The enamel organs of the deciduous teeth are just appearing

each swelling also undergoes proliferation and a small knot of compressed cellular tissue results. The epithelium of the enamel organ spills over this knot of compressed cells and encloses it on all sides except its deep aspect. Prior to the envelopment of the connective tissue knot the enamel organ is said to be at the cap stage and when the connective tissue has been enclosed within the epithelium it is known as the bell stage of development. The dental lamina between the primitive enamel organs tends to break up and disappear but the enamel organ remains joined for the present to the oral epithelium by a cord of epithelial cells. Posteriorly the dental lamina continues to grow backwards and will eventually form the tooth germs for the permanent molar teeth.

Within each enamel organ differentiation of the cells now occurs. This differentiation produces four different types of cell. The cells which lie in contact with the connective tissue in the inner aspect of the enamel organ become cuboidal and are known as the internal enamel epithelium. The cells in contact with the connective tissue on the outer aspect of the 'bell' are flatter cells and are known as the outer enamel epithelium. Between the inner and outer enamel epithelia, the cells in the interior of the enamel organ accumulate an intercellular fluid rich in mucopolysaccharides and become separated from each other apart from thin connecting cytoplasmic processes. They resemble interconnected stars and have the Latin term for a starry network applied to them, the stellate reticulum.

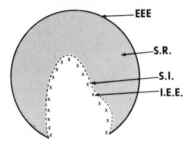

**Fig. 8.11** Diagram of the late bell stage when differentiation of the layers in the enamel organ is complete. EEE = external enamel epithelium; SR = stellate reticulum; SI = stratum intermedium; IEE = internal enamel epithelium

Between the stellate reticulum and the internal enamel epithelium a layer of flattened cells forms the stratum intermedium (Fig. 8.11). The stellate reticulum appears to provide a pressure resistant material to counteract the force of dividing cells within the enclosed connective tissue mass which is now known as the dental papilla. The cells of the internal enamel epithelium initially map out the shape of the future enamel-dentine junction by proliferating around the rim of the bell and growing deeper into the mesoderm. Cusps and fissures have their counterparts at the amelo-dentinal junction and are initiated by localised areas of increased cell division in the internal enamel epithelium. This increased proliferation causes a buckling of the epithelium into the stellate reticulum which is of the correct turgidity to allow this to happen without distortion of the whole crown pattern. The enamel organs just prior to hard tissue formation are spherical irrespective of the final shape of the tooth, an indication of the pressure within the stellate reticulum. Once hard tissue formation has started, the stellate reticulum rapidly collapses in that region and the outer enamel epithelium lies much closer to the forming enamel.

The connective tissue mass 'inside' the enamel organ is the dental papilla. At the rim of the 'bell' this is continuous with a layer of connective tissue round the outside of the enamel organ, the dental follicle. The dental follicle is responsible for the nutrition of the enamel organ, holds it in position by its connection with the lamina propria of the mucosa, and eventually will form the periodontal membrane. The cells of the dental papilla are thought to originate from the neural crest area and are sometimes referred to as ectomesenchyme. Growth of the enamel organ is matched by proliferation of the dental papilla cells. Whilst the enamel organs for the deciduous teeth are growing and differentiating the dental lamina continues to grow backwards. It gives rise to the enamel organ for the first permanent molar at about sixteen weeks of intra-uterine life and those for the second and third molars after birth. The epithelial connections of the enamel organs with the surface become thinner and eventually break down. The remnants of this epithelium occasionally proliferate and form little whorls of epithelium known as the glands of Serres. They may hinder the path of eruption of the teeth by forming cysts.

### Hard tissue formation

The internal enamel epithelium has several functions one of which we have mentioned, the mapping out of the shape of the crown. Before the full extent of the crown has been reached, the cells of the internal enamel epithelium at the tip of the crown elongate and exert an influence over the peripheral cells of the dental papilla. The dental papilla cells differentiate forming a single layer of columnar odontoblasts immediately adjacent to the enamel epithelium. These cells will form the dentine. This process of change in the internal enamel epithelium and differentiation of the odontoblasts spreads down the slope of the future crown of the tooth, whilst around the bottom of the enamel organ, the junction between inner and outer epithelium continues to grow until the full extent of the coronal part of the tooth is reached.

penetrates the gum. This breach is protected by two mechanisms, the attachment of the gingival fibres to the root of the tooth and the epithelium of attachment. The epithelium is continuous with the epithelium lining the gingival crevice and is sometimes called the junctional epithelium. It is derived from the reduced enamel epithelium which covers the crown of the erupting tooth (Fig. 5.7), though proliferation and mixing of its cells with those of the adjacent gingival epithelium make it difficult to differentiate between the two. The cells produced by this proliferation are shed into the gingival crevice. The junctional epithelium at its connective tissue interface has a basal lamina with hemidesmosomes just as any other epithelial connective tissue junction. On the enamel side there is also a basal lamina between the cells and the enamel and the cell membranes bear hemidesmosomes related to this basal lamina.

The epithelium of attachment extends coronally from the cemento-enamel junction for about 2 mm and with age it tends to migrate down the root of the tooth. It is debatable whether this occurs in the absence of pathology but it is a common occurrence. A basal lamina is then found between the cells and the cementum.

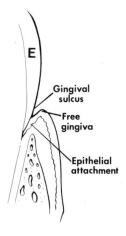

**Fig. 5.7** Diagram of section through the gingival sulcus. In a histological section the enamel (E) would have been lost in the preparation of the section and the gingival sulcus would appear wider than it normally is. The epithelium of attachment is shown here as extending onto the cementum and with age it may be found entirely on this tissue

## Clinical considerations

By the presence of a thin zone of a fibrous, vascular, innervated tissue between tooth and bone, the forces of mastication can be cushioned and controlled. The downward force is transferred to the alveolar bone by the oblique fibres and the hydraulic effect of the tissue fluid protects the fibres from damage. The nerve endings in the periodontal membrane respond and set up inhibitory reflexes to the muscles of mastication. These reflexes together with the information from sensations arising in the oral mucosa, the muscles and temporomandibular joint play a role in the chewing cycle. Reconstruction of the socket during growth by the action of the formative cells of the periodontal membrane allows teeth to migrate, for example in mesial and occlusal drift. The orthodontist makes use of this ability when he exerts force upon the teeth to move them in the correction of malocclusion.

The attachment of the periodontal membrane fibres is not the major factor to be overcome in extracting a tooth. They are readily broken down if the necessary force is applied correctly. It is the close adaptation of the socket to the shape of the root which presents the main retention. Although dilation of the socket is often necessary during extraction of teeth, excessive and uncontrolled force can lead to fracture and loss of part of the alveolar bone.

Infections in the pulp of a tooth can only spread outwards into the apical region and an apical abscess is a common sequel to pulpitis. This may show up as a break in the lamina dura on radiographs as the inflammatory process destroys the bone of the socket. In the upper jaw such infections can spread to the facial surface of the maxilla. From the lateral incisor and the palatal root of the first molar infections tend to spread toward the palate. Occasionally the antrum may be involved from the second premolar and first molar as the roots of these teeth are closely related to it. In the lower jaw an infection from the pulp or periodontal membrane may penetrate the buccal or lingual alveolar plate, depending on the relationship of the root to the bone. If an abscess points lingually, pus will collect above the mylohyoid if the anterior teeth are involved and below the mylohyoid if the apex of the tooth lies below the mylohyoid line. These infections are dangerous as potential spread to the fascial planes around the larynx is threatened.

On the buccal side infections from either upper or lower jaw may reach the superficial loose connective tissue of the cheek and spread widely. The more usual occurrence is for the pus to form a localised swelling of the gum—the so-called gum boil.

The seal between oral epithelium and the tooth is easily damaged and infections sometimes reach the periodontal membrane by this route. The alveolar bone may then be resorbed and as the area of tooth attachment is reduced, so the forces of mastication may become too great for the remaining periodontal membrane. Mobility of the tooth represent some form of systemic disturbance. Thus teeth from opposite sides of the same person will show similar sequences of incremental lines and the lines in fact can be used as for identification if two teeth of the same person are being compared. Since the teeth develop at different times it is obvious that it is not the position of the lines in the enamel which are similar but the spacing and relative density of the lines. In deciduous teeth, which are all in the process of calcifying at birth, there is a disturbance in nutrition as the child is

**Fig. 8.14** Formation of enamel. As the ameloblasts retreat they lay down the rods of enamel. At the boundaries of the rods the crystallites change direction

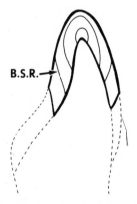

B.S.R.

**Fig. 8.15** The formation of the brown striae. In this diagram enamel formation has been halted just at an incremental line. The broken line indicates the eventual crown form

born. This is reflected in a well marked incremental line known as the neo-natal line. The presence of such a line indicates that some enamel was laid down after birth. From Fig. 8.16 it can be seen that in longitudinal ground sections the brown striae reach the surface along the sides of the tooth. On surface view these lines represent outcroppings of the prisms of enamel. They can be seen in scanning electron microscope pictures as lines running across the tooth particularly on a newly erupted tooth. They are known as perikymata and they are soon worn off with normal attrition, except in the protected cervical region (Fig. 8.17).

**Fig. 8.16** Where the brown striae meet the surface low ridges called perikymata are produced

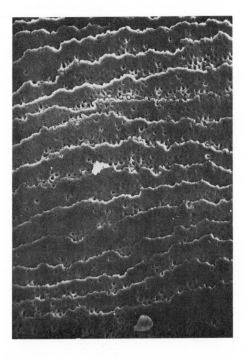

**Fig. 8.17** Scanning electron micrograph of the enamel in the cervical region. The irregular lines running across the enamel are the outcrops of the brown striae

damaging this important nerve. The excretory duct of the parotid gland which was described by Stenson and bears his name, runs forwards across the masseter muscle before turning inward at the anterior border of the muscle to pierce the buccinator muscle in the cheek. Its opening in the mouth is opposite the upper second premolar. In the living subject the duct may be felt lying on the masseter just below the zygomatic arch.

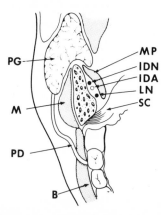

**Fig. 6.1** Horizontal section through the ascending ramus of the mandible and parotid gland. PG = parotid gland; M = masseter; PD = parotid duct; B = buccinator; MP = medial pterygoid; IDN, IDA = inferior dental nerve and inferior dental artery; LN = lingual nerve; SC = superior constrictor

Histologically the gland is made up of spherical or elongated tubular acini which are entirely serous. Thus parotid secretion has a thin watery consistency. The acinar cells have rounded or oval shaped nuclei situated close to the outside or basal part of the cell. The profiles seen on histological section vary from circular to elongated ovoid. In the centre of the acini is the beginning of the duct system and this consists of intercalated ducts, striated ducts and interlobular or

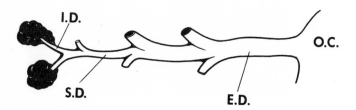

**Fig. 6.2** Diagram of duct system in the salivary gland. ID = intercalated duct; SD = striated duct; ED = excretory duct; OC = oral cavity. The excretory duct is at first interlobular then extraglandular

excretory ducts (Fig. 6.2). Intercalated ducts get this name because they are intercalated or placed between the acini and striated ducts. Striated ducts are so called because the arrangement of mitochondria aligned in rows, together with the folded cell membrane give the base of the duct cells a striated appearance in sections, under the light microscope. These striated ducts are an important part of the gland since the saliva is modified by absorption of some ions here and the secretion of others. The interlobular ducts are lined by columnar or cubical epithelium which has occasional goblet cells. At the oral end of the duct the epithelium becomes stratified and then squamous in character. The opening is guarded by a weak sphincter muscle. This can be dilated using a graded series of silver probes in investigations of the duct system in patients with salivary disorders.

## The submandibular gland

The submandibular gland lies partly on top and partly below the mylohyoid muscle. The two parts are continuous around the posterior border of the muscle. The upper part which is nearer the oral mucous membrane, lies between the mylohyoid laterally and the hyoglossus muscle medially. The deep or lower part of the gland lies between the mylohyoid muscle and the inner aspect of the mandible. A shallow depression on the medial side of the horizontal ramus of the mandible accommodates it and lymph nodes are often embedded in its surface. If one of these lymph nodes becomes enlarged and painful in infections of the teeth or gums it can be palpated with the fingers as a small solid movable mass. The facial artery crosses superficially over this lower part of the gland just before turning over the lower border of the mandible to reach the face. This is a convenient place to monitor the pulse beat particularly during a general anaesthetic in the dental surgery. The duct of the gland runs forward from the anterior border of the upper part raising a fold of mucous membrane in the floor of the mouth and opens on a little papilla to the side of the lingual frenum just behind the lower incisor teeth. It is not uncommon for stones to be found in this duct and it is possibe sometimes to milk small stones from it. The lingual nerve crosses the floor of the mouth under the duct to reach the tongue. Hanging from the lingual nerve at about the middle of its traverse of the floor is the submandibular ganglion, a relay station for the parasympathetic supply to the gland.

Histologically the gland contains acini, 75 per cent of which are serous and the remainder mucus secreting. These are arranged either in separate acini where all the cells secrete mucous or serous saliva or as a central acinus of mucus secreting cells capped by a segment of serous secreting units. These segments appear as small caps partly

## Formation of cementum

While the epithelial sheath is growing downwards and root dentine is forming, the coronal part of the sheath breaks up into a network rather than a tube. This network appears on sections as little islands of epithelial cells. Cemontoblasts now differentiate from the follicular tissue around the root and insinuate themselves between the sheath and the dentine. Cementum is now laid down on the root surface incorporating fibres from the follicle which gains an attachment to the root. In the cervical two-thirds of the root the cementum laid down during eruption is acellular.

The epithelial cell rests which appear when the Hertwig's sheath breaks up remain into adult life and can be seen on sections of the periodontal membrane. They lie slightly away from the cementum surface within the bundles of collagen fibres. They are the epithelial cell rests of Malassez and may proliferate to line cyst cavities when such lesions occur in the periodontal membrane. Their function in normal tissue is entirely unknown although it has been shown that they are alive and undergoing enzymatic activity.

## The development of the permanent teeth

At approximately the bell stage of development of the deciduous teeth there arises on the lingual side of the enamel organ a further extension of the dental lamina. This is the *anlage* for the permanent tooth. These little tooth germs remain quiescent for some time and when the deciduous teeth have erupted they develop into the crowns of the permanent teeth following the pattern of development of the deciduous teeth. In the case of the incisors and canines the permanent tooth germs remain on the lingual side and on a dried skull there are small openings in the bone which lead down into the crypts or spaces for the permanent teeth. These are known as the gubernacular canals and they house fibrous tissue which runs from the oral mucous membrane down and around each developing tooth germ. In the case of the deciduous molars the premolars develop initially on the lingual side but soon come to lie between the roots of the deciduous teeth which are widely spaced. Thus each premolar will be directly below its preceding deciduous molar. Occasionally in extracting deciduous molar teeth the premolar tooth comes with it. A replaced tooth germ may survive but often the loss of blood supply is sufficient to cause its death.

## Malformation of tooth germs

Various forms of malformation of the tooth germs can occur and altogether they are known as odontomes. A tooth germ may partially divide giving a rather wider tooth than normal or two teeth can fuse at the developmental stage. If the internal enamel epithelium proliferates too fast in one area it can grow into the papilla and an invaginated odontome is the result. It may also grow outwards producing a malformed large crown. On the root sometimes the epithelial sheath develops a stellate reticulum and enamel is produced. This is known as an enamel pearl. Parts of the dental lamina occasionally do not develop, producing partial anodontia, i.e., lack of some units in the dentition. In other cases extra teeth are formed which may or may not be like normal teeth. They are called supernumerary teeth. Extension of the dental lamina backwards has been known in rare cases to produce a fourth or even a fifth molar. Tumours sometimes arise from remnants of the dental lamina and these may contain enamel and dentine.

bacteria, epithelial cells and food debris. The bacteria can affect the composition by their metabolism and as the pH also changes on standing, the time between collection and testing can affect the result of analysis. Quantitive measurements of the constituents of saliva are rarely used clinically in diagnosis because of the wide range in values. It is worth noting that saliva, in common with other biological fluids, contains a large number of substancés in trace amounts and these have not been estimated.

Water is the main component of saliva in terms of amount present. It acts as a solvent for some of the foods introduced into the mouth and as a diluent to help protect the mucosa from harmful materials such as alcohol. Substances can only be tasted in solution and the water in saliva enables us to appreciate food materials not already in solution.

The main proteins in saliva are the glycoproteins. They have a complex structure with the proportions of protein and carbohydrate in the molecules varying from person to person. They have as their main functional characteristics lubrication and high viscosity. Saliva contains a mixture of these glycoproteins, which were originally called 'mucins' giving rise to 'mucus' in solution. The bolus of food mixed with the glycoprotein becomes a slippery mass easily propelled into and along the oesophagus. Speech is also facilitated by the free movement of the tongue on lips, teeth and palate when all surfaces are liberally coated with the glycoproteins of saliva. Lack of this lubrication gives rise to a 'dry' feeling in the mouth and probably helps in the thirst reflex. The effect of nervousness or anxiey in reducing flow of saliva is appreciated by the feeling of dryness in the mouth and throat. Difficulty in speaking is one result and the provision of drinking water for public speakers is evidence of this function of saliva. Lubrication of the mucosa aids the retention of dentures and helps to protect the mucosa against friction. The glycoproteins in saliva have the property of adsorption to all surfaces in the oral cavity. The coating on the mucosa protects it against the ingress of materials from the saliva. It has been shown in animals for example that carcinogens penetrate the oral epithelium much more easily when salivation is hindered by drugs or surgical intervention. On the teeth the glycoproteins of saliva are associated with pellicle formation, the forerunner of plaque.

The attachment of the pellicle to the tooth is thought to give a foothold for bacterial accumulation. When the pellicle has been colonised in this way it is referred to as plaque. The bacteria in plaque may produce acids by their metabolism of sugars, and the carious process may be initiated. Nearer the gingival margin the plaque bacteria produce toxins which may penetrate the epithelium and initiate gingivitis. Thus the removal of plaque is a preventive measure in both caries and periodontal disease.

Immunoglobulins in secretions from the submandibular and parotid glands are thought to be synthetised in the glands and not diffusion products from the serum. IgA, one of the immunoglobulins in blood is found in relatively high concentration in saliva. The contribution from the gingival crevices is small and edentulous patients show no reduction in salivary immunoglobulin levels. The concept of immunisation against the organisms causing caries has recently been under investigation and the rationale behind this is related to the presence of immunoglobulin antibodies in saliva.

Blood group substances in saliva are of importance in forensic science and in some genetic studies. The agglutinogens A, B and O are found in the saliva of approximately 80 per cent of the population. They are secreted mostly by mucous cells, and in secretors the activity of these substances in saliva is much higher than that of the red blood cells. The substances adsorb readily to tooth surfaces and as they are similar in molecular structure to bacterial coatings they may compete with bacteria for adsorption sites and thus inhibit the attachment of organisms to enamel.

Saliva also contains a protein known as lysozyme which is an antibacterial enzyme. Its activity is not directed equally against all bacteria and it seems to be inhibited by substances in saliva. Other antibacterial factors have been described in saliva though a full scientific evaluation of these components has not yet been carried out.

The only digestive enzyme of any significance in saliva is amylase which is found mainly in parotid and submandibular secretions. It breaks down starch to simpler sugars that are more soluble than the parent starch molecule. Since it acts optimally at pH 7–8 its action in digestion of the food is limited by the small time available while the bolus is in the mouth and the oesophagus, prior to entry to the less favourable acidic conditions within the stomach. Food particles left around the teeth and on the mucosa after a meal are the more likely targets of the salivary amylase.

The inorganic ions, sodium, potassium chloride, calcium phosphate, carbonate and hydrogen are all found in saliva, in amounts which vary with flow rates (Fig. 6.4). The variation in concentration is related to the time that saliva spends in contact with the striated duct cells as some ions are selectively reabsorbed and others are secreted by this part of the duct system. In addition water is absorbed by the duct cells at low rates of flow.

Saliva is hypotonic with respect to plasma. This hypotonicity does not affect the cells of the surface epithelium of the mouth, probably symphysis, in which there are remnants of Meckel's cartilage and small islands of secondary cartilage. The condyle is on a level with the upper border of the horizontal ramus, the coronoid process is poorly developed and the angle of the mandible is markedly obtuse (Fig. 9.1). The interior of the bone is almost all occupied by developing teeth.

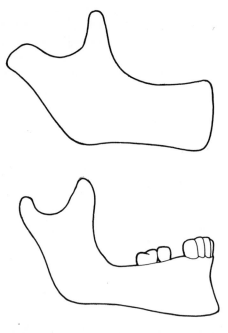

**Fig. 9.1** Outlines of the mandible at birth and in the adult. The younger bone has been enlarged to the same length as the adult bone to illustrate the changes in proportion of the various parts

By the age of twelve months the two halves of the bone have united and all trace of the cartilage has disappeared from the symphyseal region. Growth in width of the mandible will then involve only surface deposition on the facial surface, combined with resorption on the lingual surface. As the teeth erupt, bone is added to the upper border of the horizontal ramus creating the alveolar bone. Small localised areas of secondary cartilage may appear in this region but normally they are transitory. Height is added to the vertical ramus of the mandible by endochondral growth at the condylar head. This cartilaginous growth area appeared during fetal life and is now very active in contributing to the growth of the mandible. The direction of its growth is upwards, outwards and backwards. The cartilage cells do not form well orientated columns as they do in long bones. Instead they are densely packed and have a multi-directional capacity for growth and remodelling, in response to the growth of the skull, and the downward displacement of the mandible when the teeth erupt. The cartilage of the condyle therefore not only contributes height, but also width and length to the mandible. Concomitant with this cartilaginous activity surface deposition at the posterior border and lateral surface of the ascending ramus maintains the shape of the ramus and allows for resorption on its anterior border and lingual surface. The resorption at the anterior border of the ascending ramus adds length to the dental arch but in young adolescents the back of the dental arch passes inside the ramus. Room for the wisdom teeth requires an expansion of the lingual plate.

With the enlargement of the muscles of mastication the angular region of the mandible increases in surface area. It provides attachment for the masseter and medial pterygoid. The coronoid process also becomes more extensive growing in a vertical direction though its anterior and posterior borders are remodelled in line with the remainder of the ascending ramus.

Very little surface deposition occurs on the lower border of the mandible, but at about puberty there is a strengthening of the symphyseal region at its lower facial border. This gives increased prominence to the chin and can markedly alter the contour of the face.

As the mandible increases in width by the deposition of bone on the outer surface and resorption lingually, any erupted teeth must move through the bone (Fig. 9.2). The tooth socket wall is resorbed on the

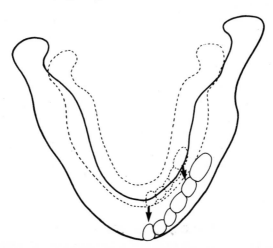

**Fig. 9.2** Outline of the mandible at three years old (in broken line), superimposed on outline of mandible at about seven years. The teeth move through the bone forwards and laterally during this period

mouth stimulates other taste buds and allows the 'adapted' organs to recover. A copious flow occurs with mechanical stimulation of the oral structures and the action of chewing alone causes about a five fold increase over the resting rate, even where the material being chewed is quite tasteless. The presence of objects in the mouth also stimulates salivary flow and dental instrumentation can lead to an embarassingly high rate of flow during dental treatment. Hence the need for salivary ejectors and other aids to keep the operation site dry. Psychic stimuli such as the thought of food may induce salivary flow and emotions such as fear or apprehension tend to reduce secretion of saliva producing a dry and 'thick' sensation in the mouth.

Measurement of salivary flow rates may be useful in assessing the severity of disease of the glands or in providing information in a course of treatment. Whole saliva can be collected by allowing the saliva to drain into a collecting cup or by spitting out at intervals making sure that no swallowing occurs. The latter method does cause some mechanical stimulation and also draws attention to salivation, which may in itself cause stimulation. It is a useful method for measuring the stimulated salivary flow rate. To measure the flow rate of the individual major glands cannulation of the salivary duct or the use of small collecting chambers which fit over the duct openings may be employed.

### Dental disease and saliva

There have been many attempts to correlate aspects of salivation with dental disease. Since caries is the most prevalent dental disease, especially among children and young adults, this disease has attracted most attention. Caries rate is measured by the number of decayed, missing and filled teeth (DMF index); but the problem is that this reflects *past* disease. Salivary studies reflect *present* conditions. Qualities of saliva that have been examined for a possible role in caries production include rate of flow, calcium and phosphate content, opsonin activity, antibacterial activity in general and against lactobacilli and streptococci in particular, and amylase activity. Only rate of flow and activity against lactobacilli or streptococci show a consistent relationship to the incidence of caries. Caries free individuals, who are rare in westernised societies, tend to have higher flow rates and tend to have smaller number of lactobacilli acidophilus or streptococcus mutans in the saliva. In patients with xerostomia caries of the cervical margins of the teeth is common.

At the present time neither periodontal disease nor the rate at which calculus accumulates has been proven to be related to any aspects of salivation.

# The functional histology of the oral mucosa

In Chapter 2 the structure of the mouth was described from a macroscopic viewpoint. In this chapter the microscopic appearance of the oral mucosa will be considered together with some of the functions of this tissue. The condition of the mucosa is a useful indicator of the state of health of an individual and an understanding of normal function and structure means that disease processes may be dealt with in a rational way.

The oral mucosa consists of covering epithelium and the underlying connective tissue which is known as the lamina propria. For descriptive purposes these are dealt with separately here but it should be remembered that the two tissues are interdependent and the structure of one affects the nature of the other.

### The oral epithelium

The oral epithelium is a stratified squamous layer which is nonkeratinised over most of the oral cavity. The areas that are keratinised are the hard palate, the gingivae and the dorsal surface of the tongue (Fig. 7.1). On the hard palate and the gingivae the mucosa is tightly bound down to the underlying bone, and on the dorsal surface of the tongue to the underlying muscle.

The basal cells are separated from the connective tissue by a basement membrane, and by their proliferation they replace the cells that are lost from the surface. The rate of proliferation of the basal epithelial cells in the mouth is higher than in skin. It has been estimated from animal studies that the oral epithelium can be replaced by division of the basal cells in about ten days. The nonkeratinising epithelium has a slightly higher rate of replacement than the keratinising variety. One important feature of normal epithelium is the property of cell adhesion. The cells remain in contact with each other and when viewed with the electron microscope appear to have small attachment sites called desmosomes between the cells. If the epithelial cells undergo neoplastic or cancerous changes they may lose their attachments and break away from the remainder of the cells.

## DEVELOPMENT OF THE DENTITION

### Deciduous Teeth

$$\frac{A\,|\,A}{A\,|\,A}$$ central incisors 6–7 months

$$\frac{B\,|\,B}{B\,|\,B}$$ lateral incisors 8–9 months

$$\frac{D\,|\,D}{D\,|\,D}$$ first molars 12–15 months

$$\frac{C\,|\,C}{C\,|\,C}$$ canines 16–18 months

$$\frac{E\,|\,E}{E\,|\,E}$$ second molars 20–30 months

### Permanent Teeth

$$\frac{21\,|\,12}{21\,|\,12}$$ incisors 6–9 years

$$\frac{3\,|\,3}{3\,|\,3}$$ canines 9–12 years

$$\frac{54\,|\,45}{54\,|\,45}$$ premolars 10–12 years

$$\frac{6\,|\,6}{6\,|\,6}$$ first molars 5–7 years

$$\frac{7\,|\,7}{7\,|\,7}$$ second molars 11–13 years

$$\frac{8\,|\,8}{8\,|\,8}$$ third molars 17–25 years

**Fig. 9:4** Eruption times for lower anterior teeth are usually in advance of the uppers. Females have slightly earlier eruption times than males. It should be noted that there is considerable individual variation in the dates at which the teeth appear. Roots of deciduous teeth take about one to one-and-a-half years and permanent teeth about two to three years after eruption to completion

permanent teeth will take their place. The final stage is the full permanent dentition. The eruption of a tooth is a process which occurs about fifty times in every individual, yet the motive force producing the movement of the tooth is not known. During eruption the tooth moves from its developmental position, through bone or soft tissue, and eventually breaks through the covering epithelium of the gum without shedding any blood. As it erupts the tooth root grows to about two-thirds of its final length, bone covering its crypt, if present, is resorbed and the epithelial covering of the crown merges with the epithelium of the gingiva. Root growth has already been described in the previous chapter. Only the permanent teeth have crypts with bony roofs, the deciduous teeth lie in shallow troughs before eruption with only soft tissue between them and the gum. Resorption of the bony roof is the result of osteoclastic activity probably induced by the pressure of the erupting tooth.

The covering of the crown of the erupting tooth is derived from the enamel organ. The external enamel epithelium collapses onto the stratum intermedium with loss of the stellate reticulum during enamel formation. Now these remaining cell layers together with the internal enamel epithelium form the reduced enamel epithelium. As the tooth approaches the gingiva the basal cells of its epithelium proliferate and grow down to meet the reduced enamel epithelium. Fusion of the two epithelia occurs and the gingival epithelium grows down along the sides of the crown until the neck of the tooth is reached. The overlying epithelium then breaks down and the crown of the tooth appears in the mouth having traversed an epithelial lined pathway in the last part of its course. The crest of the fused gingival and reduced enamel epithelia now moves down the side of the tooth exposing more and more of the crown (Fig. 9.5). This fusion is so intimate that it is difficult to differentiate between the two tissues. Gradually the reduced enamel epithelium is replaced by the proliferation of gingival epithelium which then becomes the epithelium of attachment. The rate at which this happens may vary round about a tooth, and interdentally the reduced enamel epithelium may persist for some time. Clinically this is a weak area as the reduced enamel epithelium is thinner than the oral epithelium. Thus the high incidence of periodontal pocketing between teeth may have a developmental explanation.

The first symptom of teeth beginning to emerge is the desire of the child to bite against something hard, either a rusk, teething ring or even the side of the cot. Most authorities believe that eruption is not accompanied by pain but it is often difficult to convince the mother that this is the case. There is probably some form of irritation in the gums prior to the emergence of the tooth but children rarely complain of pain as the permanent teeth erupt. Within the mouth as the tooth erupts the overlying tissues become avascular, this white area heralding the arrival of the tooth. Occasionally an eruption cyst develops over the emerging crown. This is generally darkish blue in colour and may burst spontaneously in the mouth. It is caused by the accumulation of cystic fluid in epithelial cell remnants of either the enamel organ or the original dental lamina. Rarely it may require surgical intervention to help the tooth erupt. There are several theories to account for the force which produces the movement of the tooth through the tissue. These are discussed in Chapter 12.

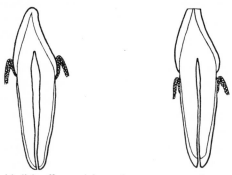

**Fig. 9.5** The epithelial cuff around the tooth exposes more and more of the crown as eruption proceeds

## Shedding of the deciduous teeth

Shedding of the deciduous teeth is preceded by the resorption of their roots and the alveolar bone. The resorption occurs by the action of giant cells, which, although they are called osteoclasts, resorb dentine and cementum, in the same way as they resorb bone. The resorptive process begins on the lingual side of the anterior teeth, where the permanent teeth are developing. In the molar regions resorption starts between the roots. The resorption is not a continuous process, and in resting phases there may be attempts at repair by the deposition of cementum on the resorbed dentine surface. The tooth may in fact become tight after showing signs of mobility. As the tooth root is resorbed the part remaining becomes less and less able to withstand the forces of mastication. The periodontal membrane becomes stressed and resorption is enhanced until the tooth is lost. Occasionally the repair process mentioned above results in a union of cementum and alveolar bone. This is known as ankylosis and it can lead to retained and submerged deciduous teeth. Submergence occurs as the growth of the alveolar margin carries the adjacent teeth in an occlusal direction. If the permanent successor is absent resorption of the deciduous root is delayed.

## Space for the eruption of the permanent teeth

There are thirty-two teeth in the permanent dentition and only twenty deciduous teeth. In addition the members of the permanent dentition with the exception of the premolars are larger than their deciduous predecessors. Extra space must therefore be created for them to erupt when the deciduous teeth are shed.

The extra space is gained by:

1. Surface deposition of bone on the facial aspects of the maxilla and mandible.

2. Growth in the tuberosity region of the maxilla to lengthen the arch.

3. Resorption of the anterior border of the ascending ramus of the mandible allows a backward growth of the lower arch.

4. The decrease in width mesio-distally in the premolars compared with the deciduous molars.

5. A greater proclination of the permanent anteriors compared with the deciduous anteriors. This means that the permanent crowns occupy a segment of a curve of greater radius than do the deciduous teeth.

In spite of this increase in the amount of room, impaction of the third molars is very common among young adults. The severity of impaction depends on the amount of space available for eruption, the inclination of the third molar tooth and the stage of development when impaction occurred. The lower third molar is more commonly affected.

It is normal to find the upper molars developing in their crypts with the occlusal surface facing distally and downwards. The lower molars develop in their crypts with the occlusal surface inclined mesially. In both jaws eruption of the molars involves a rotational movement, in addition to the movement in an occlusal direction.

It can be seen that there are many factors involved in the establishment of a normal relationship between the jaws. A close harmony in growth rates of mandible and maxilla, the timing of eruption of the teeth so that space will be available for them, the correct proportionate relationship between the size of the teeth and the size of the jaws are all important, in addition to the correct nutritional, hormonal and genetic influences which govern the growth processes.

# 10

# Calcium and phosphate metabolism

Teeth and bones require an adequate supply of calcium and phosphate to mineralise properly. An adequate supply of these minerals depends on sufficient quantities in the diet, proper absorption from the gut, utilisation at the required site and a control system to ensure that the correct levels in the blood are maintained. A knowledge of these factors is necessary to understand not only the normal metabolism of calcium and phosphate but the rationale behind the treatment of any of these mechanisms if they go wrong.

### Calcium and phosphate in the diet

According to a World Health Organisation Report in 1962 the daily requirement of calcium in the diet for the average adult person is about 0.5 gm. However this is not an absolute figure since there are several factors which may influence this requirement. During growth, for example, when bone is being formed and teeth are calcifying the daily requirement is much higher. Similarly, women require a higher amount of calcium in the second half of pregnancy and it is usual for medical practitioners to prescribe calcium tablets for mothers-to-be. Whether such supplements are strictly necessary is open to question as experiments to measure requirements are difficult to carry out on a large scale in humans. Although calcium may be withdrawn from the skeletal stores during lactation, this withdrawal is unlikely to affect the mothers' teeth. The minerals in the teeth are not readily available to the body and the old wives tale that 'every pregnancy costs a tooth' if true, must be based on factors other than the withdrawal of calcium from the teeth. Foods rich in calcium include milk and dairy produce especially cheese, some types of fish, and green vegetables. These foods are the main source of calcium in European diets.

Phosphate is present in most animal and vegetable cells and therefore is present in nearly all food. It has a high rate of turnover in the soft tissues of the body in addition to its main role in mineralisation. The daily requirement is difficult to estimate but it is probably of the same order as that of calcium. Deficiency in the diet is extremely unlikely.

## Absorption of calcium and phosphate

Calcium and phosphate are poorly absorbed as they readily form insoluble salts. Absorption of one ion favours the absorption of the other since there is less tendency then for precipitation by the combination of the two ions. The full story of calcium and phosphate absorption is complicated and all the factors involved are not yet fully understood. For convenience they can be divided into those that aid absorption and those that hinder absorption.

### Factors aiding absorption

Vitamin D is an essential element in the proper absorption of calcium. It has been known for many years that vitamin D prevents rickets, a disease in which bone and cartilage are poorly mineralised. In rachitic children the limb bones may be so severely affected that they cannot support body weight thus producing the characteristic 'knock-knees' or 'bow-legs'. Cartilage is not replaced by bone at the epiphyses and swellings occur at the ankles, wrists and sternal ends of the ribs. Vitamin D in the body comes from two sources, food and the action of sunlight on the skin. Vitamin D is fat soluble and foods rich in this vitamin are eggs, dairy produce, and the oils from some fish. Ultraviolet light acting on the skin converts 7-dehydrochloesterol to a form of vitamin D. The vitamin from both sources is acted on by the liver and kidneys to produce the active form known as di-hydroxycholecalciferol which in turn acts upon the intesinal mucosal cells, increasing their absorptive capacity for calcium.

Some hormones have been shown to increase calcium absorption. It is often difficult to separate the effects of the various hormones in the body since giving one hormone may induce the release of another. The hormones produced during pregnancy do favour calcium absorption but this may occur indirectly as parathyroid levels are raised in pregnancy. Parathyroid hormone in turn is known to be one of the factors controlling the formation of the active form of vitamin D.

The gut has the capacity to adapt to consistently low levels of calcium intake. With persons on a low intake the percentage of calcium absorbed from the diet is found to increase over long periods. The mechanism for this adaptation is thought to be under the control of the active form of vitamin D.

Factors which enhance the solubility of calcium in the gut enhance its absorption. The acidity in the stomach favours calcium absorption and patients who have had the stomach removed absorb less calcium.

Milk acidified with hydrochloric acid before drinking will have more of its calcium absorbed, and this is included in the dietary regime for gastrectomised patients. If calcium and phosphate amounts in the diet are increased, absorption will increase but the proportion of total calcium and phosphate absorbed goes down. There is therefore a limit to the beneficial effect of adding these minerals to the diet. A high protein diet has been found to increase calcium absorption presumably by the formation of soluble protein calcium complexes.

*Factors hindering absorption*
A high calcium intake over a long period of months will result in a smaller proportion of the ingested calcium being absorbed. The mechanism by which this occurs is unknown.

Lack of acid in the stomach or the upper part of the small intestine (a condition known as achlorhydria), results in conditions more favourable to the precipitation of calcium salts and therefore reduces absorption. Certain food products form insoluble salts and are thought to cause a lowering of calcium absorption. Phytates present in brown bread, reduce calcium absorption, but this effect is probably only of importance when the intake of calcium is low. It is interesting that chapatis in the diet of Asian families in Britain have been blamed for the low levels of calcium and phosphate in the serum and for the relatively high incidence of rickets in Asians in Britain. Chapatis have a high phytate content. Spinnach and rhubarb leaves contain high concentrations of oxylate compared with other vegetables. The free oxylate may react with calcium in the food to form insoluble salts which are therefore unabsorbable. Also both oxylate and phytates may form insoluble salts with the calcium in the digestive juices, preventing their reabsorption.

Phosphate is absorbed more readily than calcium but factors which affect the absorption of calcium are likely to affect phosphate in a similar way. When calcium is absorbed less precipitation of calcium phosphate occurs and hence more phosphate is available for absorption.

**Calcification of the hard tissues**
When calcium and phosphate are absorbed they enter the blood stream in amounts such that the serum appears to be supersaturated. Supersaturation of the serum does not lead to spontaneous precipitation however, as there are substances in serum which, by complexing with calcium and phosphate, can alter the activity of the ions. The main problem in the calcification of the hard tissues is to determine why calcium salts should be laid down in specific sites, i.e., the

mineralised tissues, and why it is not laid down in other areas. One might also question what restricts calcification when the required amount of hard tissue has been laid down.

Two main hypotheses attempt to explain the process. Firstly, it has been suggested that an enzyme may be present in calcifying tissues which can locally increase the amount of phosphate present in solution, thus pushing the supersaturated solution to precepitation. This enzyme, called alkaline phosphatase, frees inorganic phosphate from organic phosphates, resulting in precipitation. The enzyme is however found in sites where calcification does not occur and this hypothesis does not appear to be the whole story. Furthermore, the evidence is somewhat conflicting about which organic phosphates are involved and it is possible that other substances in the calcifying matrix induce a higher level of phosphate locally and hence cause precipitation.

The second main hypothesis suggests that there is a seeding mechanism involved. Collagen was thought to act in this way. Seeding nuclei of crystallisation once formed continue to grow as long as calcium and phosphate are available and there is enough room for calcium phosphate precipitation to occur. It is postulated that the reason why calcification does not occur in all collagenous sites is that there are differences in the nature of collagen from these sites, compared with collagen from calcifying sites. These differences are related to the size of the spaces between collagen molecules. Chondroitin sulphate, lipid substances and phospho-proteins have also been proposed as the seeds for precipitation to start.

Recently electron microscopy has shown that active odontoblasts and osteoblasts have vesicles which contain alkaline phosphatase and substances with a high affinity for calcium. These vesicles are released at the calcifying front and provide both a source of alkaline phosphatase and localised high concentrations of calcium and phosphate.

## Control of blood levels of calcium and phosphate

The levels of calcium and phosphate in the blood are affected basically by three mechanisms, absorption from the gut, withdrawal or deposit from the bone bank, and excretion by the kidney. These mechanisms are controlled by the hormones, parathormone and calcitonin. To help understand the way in which the hormones work it is useful to look at the form in which these minerals are present in blood. Calcium concentration in serum is about 10 mg/100 ml. About half is present in ionic form and the other half is bound to protein, citrate, phosphate or bicarbonate. Phosphate exists in blood as inorganic phosphate or as combinations of phosphate and sugar, phospholipids and nucleotides. Very small proportions of the total phosphate exist in the blood as inorganic phosphate. Calcium seems to be held at a more constant level in the serum than phosphate which is subject to diurnal and seasonal variation.

Parathormone is the hormone released from the parathyroid glands. It increases calcium absorption from the intestinal contents probably through the action of vitamin D. Osteoclastic resorption of bone is increased by the action of parathormone, thus raising the level of calcium and phosphate in the blood. By its action on the kidney it increases urinary phosphate and decreases urinary calcium, thus the net effect is an increase in the level of serum calcium.

Calcitonin was discovered in 1962 when the serum levels of calcium were being studied in dogs which had had their parathyroids and thyroids removed. The experiments showed that another calcium controlling hormone was released when there were high levels of calcium in the blood. This hormone was later found to exert a direct effect on bone, reducing resorption or increasing deposition or both. It may also have an effect on the gut, reducing the absorption of calcium. It does not appear to act directly on the kidney. Calcitonin is produced by cells in the thyroid, and at one time was known as thyrocalcitonin. It is also produced by cells in the parathyroid and thymus so the 'thyro' part of the name has been dropped. Calcitonin has been used therapeutically in the treatment of Paget's disease where bone resorption is abnormally high and in cases of hypercalcemia.

### Calcium and phosphate in tooth formation

The mineral element in the teeth is not easily removed and the teeth play little part in the maintenance of serum levels of calcium and phosphate. The developing dentition also seems to occupy a preferential situation in that the degree of calcification during development is not affected until comparatively low levels of calcium and phosphate in the blood are reached. These levels are lower than those which will affect bone deposition. Vitamin D deficiency produces a hypoplasia of the enamel in dogs but there is no real evidence that any hypoplastic conditions of human teeth are related to the lack of vitamin D. Both vitamin D and vitamin K have been suggested as anti-caries agents. The mechanism for vitamin D could be that vitamin D promotes a higher degree of calcification but this is doubtful. Vitamin K inhibits acid production by salivary bacteria, but further work on this protective effect is needed. Fluoride incorporation in enamel, either during the formation of the tooth or for a short time after eruption, leads to a reduction in the incidence of caries. The fluoride becomes incorporated in the lattice of the apatite crystal making it less soluble in acid. This is an important ion in this connection and for details of the utilisation of this ion the larger text books should be consulted.

## Non-skeletal utilisation of calcium and phosphate

Calcium ions are necessary for blood clotting. Rarely does this cause any bleeding problems but the fact is used in the collection of unclotted blood. Certain examinations of blood require blood that is not clotted, and for these the blood is mixed with citrate, oxylate or ethylenediaminetetraacetic acid. All of these are chelating agents and they act by removing the calcium ions from the blood. Conversely calcium hydroxide (styptic) sticks are used to promote clotting of blood from cuts on the skin, though not in the mouth. Calcium takes part in the junctional complexes between cells and the rate of mitosis of cells is affected by changes in the calcium concentration of the environment. It is an essential mineral for the normal contraction of cardiac and skeletal muscle and for transmission at nerve endings.

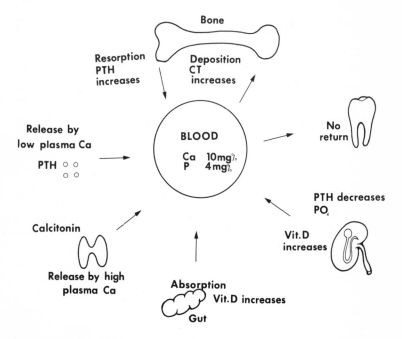

Fig. 10.1 A diagram of the main factors involved in the utilisation of calcium and phosphate by the hard tissues (modified from Jenkins's *Physiology and Biochemistry of the Mouth*). PTH = parathormone; CT = calcitonin

Tetany, an uncontrolled contraction of muscles, is caused by increased activity at motor end plates at low levels of calcium ions in the blood. This condition is seen in cases of hypocalcemia.

Phosphate is almost ubiquitous. Energy cycles within the cells and the structure of cell walls involve phosphates in combination with sugars, fats and nucleic acids.

### Excretion of calcium and phosphate

The glands of the digestive tract secrete calcium and phosphate into the intestinal canal, in the digestive juices. A proportion of the secreted material is absorbed with the products of digestion but some is lost in the faeces. In addition unabsorbed calcium and phosphate from food sources are excreted in the faeces. The total calcium and phosphate lost by this route is not as directly controllable as the excretion by the kidney.

In the kidney calcium and phosphate are reabsorbed by the cells of the tubules, from the glomerular filtrate. Control of the urine content can be exercised by increasing or decreasing the amount which is reabsorbed. The hormonal control of blood levels of calcium and phosphate is effected in this way and the amount excreted varies with the bodily requirement. For example during growth and in late pregnancy the urine content of calcium is reduced.

Some calcium is lost in sweat but the loss is dependant only on volume of sweat and not the needs of the body. Although calcium and phosphate are secreted in saliva the majority of this is reabsorbed in the gut. The main factors involved in calcium and phosphate metabolism are shown in the diagram (Fig. 10.1).

# 11

# Function in the oral cavity

Absorption of food can only take place when the food has been made soluble. The first step in this process is mastication by which the food is comminuted and lubricated with saliva. Mastication is a co-ordinated series of repeated movements of the stomatognathic (chewing) apparatus, carried out mainly under reflex control but voluntary control may be superimposed when required.

The action usually begins with incising if the food is too large to go into the mouth. The lower incisors move downwards and forwards, then upwards through the food, meeting edge to edge with the upper incisors. Pieces of food of the correct size are placed directly to either the right or left side of the mouth and chewing starts. The preferred side may be the result of habit and seems to be unrelated to right or left handedness. Local conditions such as toothache, missing teeth or mouth ulcers will also influence the side to which the food is placed. A bolus is produced when the softened food is mixed with saliva, but the pattern of chewing shows great individual variation. In some subjects the bolus is divided into two by the tongue and chewing goes on simultaneously on both sides. In others chewing is unilateral but the bolus is usually changed from one side to the other during mastication. It has been noted that when the bolus is shifted from one side to the other the pattern of chewing may be different, in that there may be more lateral deviation with one side than the other.

In the first part of this chapter the various components of the stomatognathic system will be considered and later we shall discuss the establishment of the occlusion, swallowing and speech.

**The muscles of mastication**
The principal muscles of mastication are the masseter, the temporalis, the medial and lateral pterygoids, the digastric and the mylohyoid. The masseter is attached above to the zygomatic arch and below to the angle of the mandible on its lateral aspect. It has two parts, the more superficial fibres running down and back so that it has a small forward component in its action and the other deeper part having fibres

running vertically. This is a powerful crushing muscle which can be palpated in the cheek if the teeth are clenched. The temporalis has a wide attachment to the side of the skull and its fibres converge on the coronoid process of the mandible to which it is attached on both sides. The posterior fibres help in retracting or retruding the mandible as well as acting with the more anterior fibres to bring the teeth together. The medial pterygoid muscle comes mainly from the lateral pterygoid plate of the sphenoid bone and is directed downwards, backwards and laterally to the region of the angle of the mandible on its medial aspect. Thus, the masseter and medial pterygoid muscle grasp the lower part of the ascending ramus between them, and acting together they can develop considerable crushing force. The lateral pterygoid is also attached to the lateral pterygoid plate and its fibres course backwards to the neck of the mandible and the articular disc of the temporo-mandibular joint. Acting with the digastric and the mylohyoid, it opens the mouth and acting with the opposite lateral pterygoid it protrudes the lower jaw. Unilateral activity swings the chin to the opposite side.

A knowledge of the direction of pull of these muscles is important in dealing with fracture cases. The displacement of the fragments when the jaw is broken depends on the direction and position of the fracture line and also on the muscles which are attached to the two fragments. Normally the rest position of the mandible is the equilibrium point where all the muscles are in balance. A loss of the pull of antagonists such as might occur in a fracture results in the fragments being displaced in the direction of the strongest attached muscle providing that the line of fracture allows this. For example, if the neck of the condyle is fractured the head may be pulled forwards by the lateral pterygoid, a movement which is normally opposed by the backward pull of the posterior fibres of the temporalis acting on an intact mandible. The larger fragment may well be pulled up towards the temporomandibular joint by the action of the masseter and the medial pterygoid, a movement normally prevented by the head of the condyle.

Other less powerful muscles act during mastication. The tongue plays a major role in controlling the bolus of food and helping push it between the teeth. The lip and cheek muscles, especially the buccinator, act to push the food into the space between the upper and lower teeth ready for the next chewing stroke. They also keep the sulcus clear of food.

### The temporomandibular joint
The mandible articulates with the temporal bone in a synovial joint on

either side. Thus a movement in one joint is always accompanied by movement to some degree on the opposite side, if the mandible is intact. Lying between the condyle of the mandible and the glenoid cavity of the temporal bone the temporomandibular disc (meniscus) separates the joint into upper and lower compartments.

The disc, composed of densely packed collagen fibres, is attached at its periphery to the capsule of the joint, and through the capsule to the margins of the articular surfaces. Its attachments at the lateral and medial extremities of the condylar head are particularly strong. Posteriorly the disc is much looser in structure with a good blood supply and an abundance of elastic tissue. With the capsule this posterior part is attached to the squamo-tympanic fissure and to the back of the condyle. Anteriorly the disc is attached to the lateral pterygoid muscle. Both the upper and lower surfaces of the disc conform to the shape of the contiguous bony surfaces. In front the articular eminence increases the length of the upper compartment. If the mandibular condyle moves too far down the slope it may dislocate into the temporal fossa. This is a common occurrence in dental extractions under general anaesthesia. The condylar head must be depressed and pushed backward to relocate it. The only strong ligament that tends to prevent dislocation is the lateral or triangular ligament, a dense thickening of the capsule on the lateral side.

In the adult the head of the condyle is covered with fibrous tissue. Before the end of the second decade of life the head has cartilage under its fibrous covering. This cartilage is an important growth site of the mandible and it adds to the length, height and width of the mandible until about twenty years of age. The glenoid fossa is also lined by dense fibrous tissue. In the non-articular parts of the two compartments there is a lining of synovial membrane which contributes synovial fluid to the joint cavities.

The movements of the head of the mandible in the glenoid cavity during mastication are complex but consist of two basic actions. The first is a hinge movement that takes place between the head of the condyle and the articular disc. This results in a simple up and down movement of the lower teeth. The second movement is a sliding movement of the disc on the anterior articular slope of the glenoid fossa carrying the head out of the cavity in the temporal bone. This produces a forwards and backwards movement of the lower jaw. Unilateral sliding results in lateral excursions of the mandible. During mastication the incisor teeth may be brought together by protruding the lower teeth for biting and tearing. For crushing and grinding the jaws move laterally during each stroke as well as up and down. During normal opening of the jaws there is first a rotation or hinge movement until the incisor teeth are about 3 mm apart. The condyle and disc then start to slide forward and accompany further hinge movement. Closing is the reverse of opening.

### Reflex control of chewing

Nerve endings in the temporomandibular joint, the muscles and the periodontal membrane are thought to control the action of muscles through subcortical centres. The nerve endings in the TMJ are found in the capsule, the triangular ligament and the articular disc. These are concerned with information about the position of the lower jaw. In the muscles there are muscle spindles to give information about the state of contraction and position of the muscles and also to control the rate and degee of contraction. The periodontal ligament is richly supplied with proprioceptive endings to monitor the degree of pressure on the ligament.

During mastication food in contact with the mucosa and the teeth initiates a closing movement of the jaws. Pressure upon the teeth and gums stimulates the jaw opening muscles and inhibits the closing muscles. As the jaw opens stretching of the closing muscles and impulses from the TMJ induce a jaw closing reflex and the movement is reversed. Thus a chewing cycle is set up. The details of the co-operation between all the nerve endings is complex and it is not certain whether there is an underlying rhythm generated in the CNS or whether the local reflexes control all the action. Probably a rhythmic activity is set up in the system once food in introduced, but this activity is modified by the local stimuli. The stimuli from the mucosa and tongue determine the various patterns of masticatory activity in response to the nature of the food as the process continues. Although the periodontal membrane is an important source of controlling stimuli during mastication, the oral mucosa plays a role and takes over this function of the periodontal membrane entirely in the edentulous patient. The type of food being eaten determines the rate and pressure exerted during the chewing stroke. Softer foods are chewed more quickly and with less force than harder and tougher foods. When an unexpectedly hard object is encountered between the teeth there is an immediate reflex inhibition of closing and a reflex opening jerk. When breaking a nut with the cheek teeth enough force is exerted to crack the shell but when the force is no longer resisted, i.e., when the shell breaks there is again an immediate relaxation of the closing muscles and a contraction of the opening muscles so that the teeth are protected.

### Occlusion

When the opposing teeth are brought together into a position where they are in maximal contact, this is known as centric occlusion. With the teeth in this position, the condyle is normally in its most relaxed retruded position in the glenoid cavity and this is known as the centric jaw relationship. This relationship is an important one in Prosthetics when some or all of the teeth are missing. In trying to reproduce the correct centric occlusal position the dental surgeon must first establish the centric jaw relation. In subjects with a full dentition the normal centric occlusal relations are:

1. The upper anteriors overlap the labial surfaces of the lower anteriors.

2. The upper canine occludes in the space between the lower canine and the first premolar on the labial side.

3. The buccal cusps of the upper cheek teeth lie outside the buccal cusps of the lowers.

4. The buccal cusps of the lower cheek teeth meet the fossae and fissures in the upper cheek teeth. The mesio-buccal cusp of the first upper molar meets the lower first molar on its buccal surface in the fissure between the two buccal cusps (Fig. 11.1).

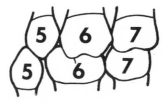

**Fig. 11.1** Arrangement of the first molars (6) in Angle's Class I relationship

The relationship of the first molar is the basis of the well known Angle's classification of occlusion. In this classification the relationship stated would represent a Class I occlusion. If the lower tooth occludes in front of this position so that the mesio-buccal cusp of the upper first molar meets the lower first molar distal to the fissure this occlusion is termed Class II. In a Class III relationship the mesio-buccal cusp of the upper molar lies in front of the fissure on the buccal surface of the lower first molar. There are further subdivisions of this classification depending on the positions of the anterior teeth. The classification is still widely used in Orthodontics though it has been extended and modified to include the relationships of teeth other than the first molars.

The teeth can, of course, articulate with each other in positions other than centric occlusion. The student may demonstrate this readily by bringing his teeth together with the lower jaw protruded or swung to the side. These positions are used to some extent during mastication. The effect of these movements during chewing is to produce areas of wear in places where the teeth come together. These have already been mentioned as wear facets and it has been pointed out that their position on an isolated tooth may help in its identification. As an exercise the position of the expected wear facet should be worked out for each tooth and compared with examples of actual teeth. It will soon become apparent that there is a considerable individual variation in occlusal patterns. In making dentures consideration should be given to the lateral and protrusive movements of the jaws, to avoid intercuspal interference during function. Such interference would lead to instability of the denture.

### The development of occlusion
The development of a 'normal' functioning occlusion is controlled by many factors. The malfunctioning or mistiming of any of these may result in malocclusions varying from a misplaced tooth to a gross disparity in the size relationships of the jaws.

The first teeth to erupt are the lower deciduous incisors followed by the upper deciduous incisors which normally are placed labial to the lowers. At about one year old the first deciduous molars erupt with the mesial surface of the lower slightly in front of the upper. The relative antero-posterior relationship of the jaws is thus established and it is further stabilised at twenty-four to thirty months when the second molars erupt. At three years the distal surfaces of the upper and lower second molars are almost in the same vertical plane. Wear on the cusps and growth of the jaws result in the lower second molar gradually moving mesially relative to the upper and by about six years old the incisor teeth meet edge to edge. When the lower first permanent molar erupts it occludes with the distal cusp of the upper second deciduous molar. The upper first permanent molar on eruption is guided into its antero-posterior position by the distal surface of the second deciduous molar and the mesial cusps of the lower first permanent molar. Its bucco-lingual position is related to the muscular forces of the tongue and cheeks. In the mixed dentition stage from seven years to twelve years canines and premolars erupt and are guided to their position by contact with the adjacent and opposing teeth. Room is made for the premolars by the disparity in size between them and the larger deciduous molars. The upper incisors are inclined labially and hence their incisal edges form a curve with a greater circumference than that formed by the deciduous incisal edges. Early loss of deciduous teeth often results in lack of space for the permanent teeth but this is especially important in the canine and premolar regions. With the loss of a deciduous molar the first permanent molar tends to move mesially, lessening the space for eruption of the premolars. By the age of thirteen years the adult occlusion is well established.

It is important to remember that this stage has been reached by the co-ordinated growth patterns in the maxilla and the surrounding bones on the one hand, and in the mandible on the other. In addition to the growth and eruption of the teeth it has also involved a balanced functional activity of the masticatory and the facial muscles.

When the teeth are lost and dentures are made to replace them it must be appreciated that the denture is a block of solid material resting on the mucosa, whereas the natural dentition consists of independently moving units, embedded in the bone. This inevitably results in new patterns of masticatory activity which a patient must learn and it is not surprising therefore that dentures do not come near the efficiency of the natural dentition in masticatory function.

## Deglutition

Deglutition or swallowing can be divided into three stages, preparatory, transitional and oesophageal stages. In the preparatory stage the bolus of food well lubricated by saliva is collected between the dorsum of the tongue and the soft palate. The teeth come together and the lips usually are closed. As the pressure on the bolus is increased suddenly it passes rapidly through the oropharynx in the transitional stage. This action is brought about by squeezing the bolus by the muscles of the tongue and those of the soft palate. The soft palate closes off the nasopharynx by the contraction of the levator and tensor palati muscles and the upper fibres of the superior constrictor. The bolus of food is received into the open end of the oesophagus, passing over the epiglottis which closes like a lid over the larynx. Simultaneously the larynx is lifted against the base of the tongue by the pharyngeal and suprahyoid muscles. This upward movement effectively closes the opening of the larynx and respiration is temporarily inhibited. When the bolus has reached the top end of the oesophagus, peristaltic waves of contraction of the muscles of the oesophagus carry it down towards the stomach, where the oesophageal sphincter relaxes to allow the food to enter the stomach.

Swallowing of liquids is very similar to the swallowing of food except that the epiglottis acts as a water shed to protect the opening into the larynx. The action of the musculature in the oesophagus can be inhibited voluntarily with practice and thus it is possible to swallow at one time large quantities of fluid.

Although the act of swallowing is reflexly controlled, and goes on automatically when eating, it can be initiated voluntarily. It is difficult to stop, once it has started. If the mouth is open wide touching the back of the mouth or tongue leads to retching rather than swallowing. This is a frequent occurrence when impressions are being taken of the upper jaw. The sensitivity to the retching reflex varies considerbly among individuals.

Note that the lips and the teeth are brought together during swallowing. In the infantile type of swallow the lips are apart and the tongue pushes between the upper and lower gum pads. This swallowing pattern sometimes is retained into childhood and may result in distortion of the position of the upper teeth relative to the lower.

## Speech

During speech the larynx produces sounds which are modified by their passage through the mouth and nose. Vowel sounds are modified by the action of the tongue, lips and cheeks, without any obstruction to the passage of air. Consonants are produced by interruption or slowing of the pasage of air by the contact of tongue with the palate or teeth or by the action of the lips. For some sounds the air is passed through the nasal cavity rather than the mouth. Cleft palate results not only in defects of speech but also of sucking and swallowing. Usually treatment of cleft palate involves close co-operation between speech therapist, surgeon and orthodontist.

The fitting of dentures to a patient may result in interference with speech patterns but usually the patient learns to adapt the action of the tongue and lips to the new situation.

# 12

# Some controversies in oral biology

There are many areas in oral biological studies where our knowledge is incomplete. In this chapter I shall present a few of the classical controversies where opinion has been divided or is still divided on the interpretation of the results of much research. Some of these are of more clinical significance than others, but clinical application is not the only criterion which inspires research workers in the dental or indeed the medical field. Knowledge is often sought for its own sake outside the constraints of clinical usefulness as we strive to learn more about ourselves and our environment. In the pursuit of such 'pure' research, information is often gained which, although not of immediate benefit, may enable other workers to progress in a more applied direction.

**Tooth eruption**

Histological studies have revealed the changes in the tissues as a tooth moves through hard and soft tissues to reach the oral cavity. The nature of the force or forces which produce these changes has not yet been elucidated. The theories to account for the force can be divided into those that suggest the force is produced inside the tooth, and those that suggest that the force is a result of activity in the surrounding tissue (Fig. 12.1).

The force acting from within the tooth may be due to the proliferation of cells within the pulp. As each cell divides it exerts a force which when accumulated could press on the only area of the tooth which does not have rigid walls, namely the open apex. Likewise the blood vessels inside the tooth may have a special arrangement of their muscular walls or activity, or the capillaries may be less permeable so that a higher vascular pressure may exist in the pulp than in the surrounding tissue. Deposition of dentine at the periphery of the pulp results in a pressure on the soft tissue since its volume is reduced. In each of these theories it is suggested that the pressure generated pushes against the apical tissues lifting the tooth. However, experiments with the continuously erupting rodent incisor, a favourite tool for workers on tooth eruption, have shown that even with the pulp removed the tooth still erupted. In addition when the proliferation of cells was reduced eruption rate was not affected and changes in blood pressure had also no effect on eruption rate.

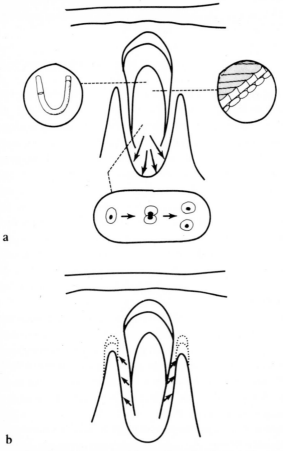

a

b

**Fig. 12.1** Illustration of the suggested forces of eruption. In (a) the force comes from the blood pressure, the proliferation of cells or from dentine production. In (b) alveolar bone growth or the periodontal membrane may provide the motive force. In (a) the tooth is pushed, in (b) the tooth is pulled through the bony socket

The forces acting outside the tooth may be due to the growth of alveolar bone in the socket of the tooth. However in some cases the tooth can be seen to erupt without any growth of the bone. The eruptive force has also been postulated as arising in the periodontal membrane. During eruption the collagen fibres from the cementum and bone form an intermediate plexus about half way across the space between them. As collagen matures it is thought to undergo contraction of the molecules and this contraction pulls the tooth upwards because of the inclination of the fibres. This theory has many drawbacks and a modified version suggests that the fibroblasts in the periodontal ligament have contractile capability. The eruptive force comes from the contraction of these cells.

The most generally accepted view is that the eruptive force is generated in the periodontal membrane but whether or not other forces are involved is uncertain.

### Sensitivity of dentine

In the dental surgery when the enamel is breached and dentine exposed, any instrumentation on the dentine produces pain if the tooth is viable. Pain is also felt on eating sweets, ice-cream or hot meals if the dentine has been exposed by caries or by gingival recession. Since dentine apparently does not have nerves running through it, it is difficult to explain its sensitivity. Attempts to show nerve fibres in the dentine have been inconclusive as many histological stains for nervous tissue are incompatible with the decalcification procedures necessary to cut sections of dentine. Electron microscopy has shown that a few nerve fibres do enter the dentinal tubules but that they penetrate only short distances. The problem of how stimuli reach the recognisable nerve endings in the pulp still remains. It is possible that the odontoblast process acts as a nerve fibre, and indeed it has been shown in lower animals that odontoblasts originate from the neural crest area. Conflicting with this possibility is a recent finding that in adults the odontoblast process does not extend to the enamel-dentine junction which is often the most sensitive part of the dentine. By the application of chemical substances to cut dentinal tubules and simultaneouly recording the nerve impulses in the sensory nerves to the tooth in the cat, evidence of a movement of fluid in the tubules has been collected. This has led to the hydrodynamic theory, postulating that the movement of fluid in the tubules somehow stimulates the nerves among the odontoblasts at the end of the tubules.

Thus, even a wisp of cotton wool drawn across the cavity floor may be enough to move fluid out of tubules and stimulate the nerve endings in the odontoblasts. Similarly heat and cold by expanding or contracting the fluid would induce movement, and osmotic effects of sugar likewise would stimulate the nerve endings. The only sensation provoked is pain irrespective of the stimulus. The more basic question of the need for this sensation in the tooth remains unanswered.

## Taste

It is possible to distinguish in the mouth four tastes, sweet, sour, bitter and salt. Two others have been described but are not so well characterised. These are alkaline and metallic. The mechanism for differentiating these tastes is not known. One possibility is that there are specific taste buds related to each of the basic tastes and that each taste bud will have a receptor for one type of chemical substance. However, recordings of the action potentials from single taste buds in animals show that some respond to all four basic tastes while others are insensitive to one taste and weakly sensitive to another. These experiments led to the pattern theory which suggests that the taste depends on the relative number of impulses in a group of nerves. For example a sweet taste bud would respond to low concentrations of sugar but only to high salt concentrations. A salt taste bud would respond to low salt concentrations but only to high sugar concentrations. The brain interprets more impulses from the first taste bud than the second as sweet, and when the impulses from the second exceed the first then salt is appreciated. Another possibility is that few substances really have a pure taste. Most contain a small proportion of other tastes besides the predominant one. This suggestion accounts for the variations there are in the basic tastes, e.g., the well known bitter taste that accompanies the sweet taste of saccharin. If the mouth is rinsed with a mixture of substances which will saturate three of the four tastes then testing with the fourth will produce a so called 'flat' taste lacking in overtones.

The mechanism whereby a chemical substance gives rise to a nerve impulse is still uncertain. The specific chemical may stimulate the taste bud by combining with a substance in the cell wall of a taste cell. This combination may be related to the shape or complexity of the molecule to be tested. It also has been postulated that the activity of some of the enzymes which are present in taste cells may be altered by sapid substances and the effect of these alterations on all cell contents might give rise to specific nerve impulses.

It should be noted also that some substances enhance the taste of other materials and are used in food manufacture, e.g., to suppress the bitter taste while enhancing the sweet taste of some foods. The sense of taste is of course complicated by the sense of smell. Most sapid substances have an effect on the olfactory area as well as on the taste buds. If air is prevented from passing over this area either by holding the nose or by mucus when a person has a cold, much of the flavour of the so-called taste is lost.

It is not known how the multitude of smells that one can detect are appreciated at the sensory cortex.

Taste is very important to the dental surgeon. Medicaments to prevent dental disease are more acceptable if they do not produce unpleasant tastes. A good example of this is chlorhexidine which reduces the amount of bacterial plaque. One of its disadvantages is the bitter taste which deters patients from regular use either as mouthwashes or toothpastes.

## Mesial drift of teeth

In spite of the attrition that goes on at the approximal contact areas, the teeth are maintained in close contact with each other. The posterior teeth drift forwards so that gaps do not open up at the front of the mouth and this is called 'mesial drift'. If a tooth is lost from the arch the posterior teeth often, though not always, tend to move forward and close the gap. There is also a tendency for the tooth posterior to the gap to tilt towards it. The forces producing mesial drift and tilting are not fully understood but are thought to be related either to the collagenous fibrous tissue in the cervical region of the periodontal membrane or to the direction of the slopes of the cusps of the opposing teeth.

In monkeys grinding the approximal surfaces of the molars resulted in accelerated mesial drift. The drift was still seen where the opposing molars had been extracted and thus occlusal forces had been removed. If the interdental fibres of the periodontal ligament were severed, grinding the approximal surfaces did not result in mesial drift. On balance it would appear that the trans-septal interdental fibrous tissue is responsible for the mesial drift force, just as the fibrous tissue of the periodontal membrane may be responsible for the erupting force.

Any theory or hypothesis to explain mesial drift must also explain the exceptions to this phenomenon. Occasionally a first permanent molar is removed and the second premolar drifts posteriorly even to the extent of contacting the second molar leaving a gap between the two premolars. More evidence is needed before we can be certain that we know all about the forces involved in mesial drift.

## Enamel maturation

When enamel is first laid down the calcium phosphate is in the form of small crystals, or crystallites. These crystallites enlarge during the process of maturation and thus the enamel matrix progresses from a soft state to form the hardest tissue in the body. The pattern of maturation has been traced by microradiography and is shown in Fig. 12.2. A thin layer nearest the amelodentinal junction matures first and the process spreads outwards to reach the outer surface at the occlusal tip of the cusp first. Maturation then spreads cervically and laterally from the amelodentinal junction. During maturation the organic matrix is reduced and the proteins are altered in the relative proportions of their amino acid make up. Water is also lost from the enamel. The mechanism whereby the calcium phosphate crystals increase in size is not known. Calcium and phosphate must be made available but how this occurs deep in the tissue remains obscure. The calcium phosphate crystals are in fact larger here than in any other hard tissue. How are the proteins and other organic materials removed? The ameloblasts apparently play some part in the removal as they can be seen to change their morphology when maturation is occurring. The removal is apparently selective but how changes in the protein are accomplished in the enamel at some distance from the cells is again unknown. There is evidence that maturation continues after eruption by the incorporation of ions from saliva.

**Fig. 12.2** The pattern of enamel maturation revealed by microradiography. The extent of the crown is shown by the broken line. The hatched area is enamel matrix and the dense black area is mature enamel. Maturation starts at the tip of the cusp, progresses both cervically and outwards. The wave of maturation is at an angle to the brown striae of Retzius

### The attachment of epithelium to enamel
There has been controversy over this topic since the early 1920's. The junction between enamel and epithelium is unique in that it is the only place where the epithelium is breached by a hard tissue to which it is not contributing. Hairs and nails do penetrate the epithelium but they are continually being formed from the epithelium.

One school of thought felt that the ameloblasts formed an organic layer over the enamel as a last product and to this the reduced enamel epithelium adhered. The other main school believed that the epithelium was held tight to the enamel by the fibres of the periodontal membrane beyond the neck of the tooth and only at the cemento-enamel junction was the epithelium attached in a line running round the tooth.

The modern view is that the epithelium is attached to the enamel by hemidesosomes to an intermediate material akin to the basal lamina of epithelial-connetive tissue junctions elsewhere. The nature of the amorphous material of the basal lamina and the nature of its attachment to the enamel have yet to be clarified. The attachment moves along the side of the tooth during eruption. With age it may move onto the cementum, a migration which some believe to be physiological. Others feel that such migration will not happen in the absence of pathological destruction of the gingival fibres just below the attachment.

## Conclusion

These few examples of the areas where our knowledge is incomplete illustrate the need for continuing dental reseach. Biology is not an exact science comparable to the physical sciences. The biological variations from tissue to tissue, from organ to organ and from individual to individual continually frustrate the scientist who wants to formulate rules to suit all reactions. Fortunately no two persons are exactly alike, it would be a dull world if we did not have biological individuality.

Finally the student should remember that any textbook at best contains only the information that was available at the time it was written. New researches or new interpretations often change concepts which were once thought to be solidly based. An open mind is the key to progress.

# Index